Passive Solar Energy

Second Edition

The Homeowner's Guide to Natural Heating and Cooling

Bruce Anderson
Malcolm Wells

Illustrations by Malcolm Wells

BRICK HOUSE PUBLISHING COMPANY
Amherst, New Hampshire

Cover photo: Douglas Balcomb

Interior view of the large solar room of Douglas and Sarah Balcomb's house in Santa Fe, New Mexico.

The publisher gratefully acknowleges the editoral assistance of Jennifer Adams in the preparation of this new edition.

Contents

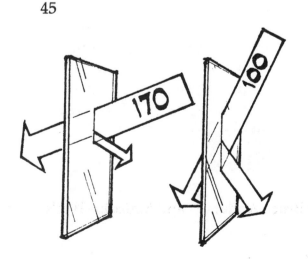

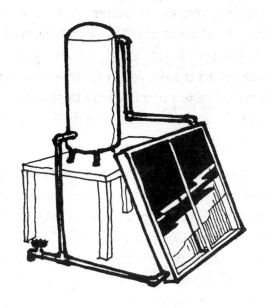

Preface

Some skeptics say energy taken directly from the sun won't work. But you know better; you've been experiencing it all your life. Think of how nice and toasty your sunporch gets. And the way you can almost bake bread inside your car on a sunny October day. Remember how your favorite begonias had sunstroke in the greenhouse last April? There's a lot of heat in all that light.

Heating and cooling with the sun doesn't have to be complicated and expensive. This book is for all of us who have burned our behinds on the sun-baked seats of our cars, who are concerned by the environmental damage being done by fossil fuels, and who are ready to do something about it all—just as soon and as simply as possible. Passive solar heating and cooling does not depend on pumps or fans or any other devices. Instead, it relies on the natural ebb and flow of the energy of the sun through a house.

With a few facts from this book and a little common sense, you can, by combining passive solar design with energy conservation, *reduce by as much as 85%* the heating and cooling bills for a new house from that for a conventional house. And many of the ideas can be adapted to existing houses as well.

It's no wonder, then, that more and more people who are planning to build houses someday are thinking passive solar. It is the natural first step toward living better while using less energy.

Our aim is to make this presentation as simple as the systems it describes. So, here it is. We hope it will be helpful.

Bruce Anderson & Malcolm Wells

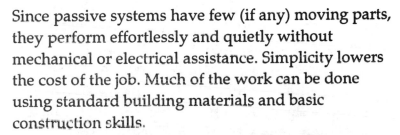

Why Passive?

Passive solar designs are simple. This simplicity means greater reliability, lower costs, and longer system lifetimes.

Since passive systems have few (if any) moving parts, they perform effortlessly and quietly without mechanical or electrical assistance. Simplicity lowers the cost of the job. Much of the work can be done using standard building materials and basic construction skills.

The most significant reason that passive makes sense economically is that most passive designs are inherently durable. Conventional building materials such as glass, concrete, and brick weather well and are generally long lasting.

For the life of the house, a passive system should continually maintain, if not improve, its value at least as well as the rest of the house. It should require little more maintenance than a standard wall or roof, and last at least as long as the rest of the house.

Because you can build passive designs in small sizes, the initial effort need not involve a large financial commitment. Instead, the first step can be relatively small with correspondingly little risk.

1

For optimum performance, some passive systems require daily or monthly adjustments of shades, shutters, or vents. Although some people may at first regard this as an imposition, it is really no more trouble than operating a dishwasher or closing draperies in the evening.

Before long, passive-home residents find these to be pleasant routines that bring them closer to the flux of the environment to which their homes are attuned. They are usually rewarded with a rich and exciting living experience as a result of their efforts, while saving both energy and money.

Radiant heat from large passive collecting surfaces is usually more comfortable than the drafty heat of conventional hot air or hot water heating. In well-designed systems, temperature variations are small, generally within a range of 5 to 10°F each day. But in less well designed houses, temperatures can vary more widely. Some solar enthusiasts feel such temperature fluctuations are natural, and not uncomfortable, particularly at the higher end. In fact, many passive solar home residents enjoy the warmer-than-usual temperatures on a sunny winter day.

Is this really true or is it just a lot of passive sales talk?

Passive solar systems help conserve fossil fuels. And since passive energy systems do not require transmission lines, pipe lines, or strip mines, they produce neither dangerous radioactive wastes nor polluted air and water. Passive systems have few negative consequences. They can use renewable and recyclable materials.

If a house has low heating or cooling requirements, and if a passive solar system is designed to provide only a small fraction of the energy, the system can be small and have only a slight effect on the overall appearance of a house. In fact, properly designed passive houses are more beautiful than conventional ones. Picture it—large expanses of glass facing south, overlooking your yard; a beautiful sunspace filled with plants year 'round. You can save energy, save money, and provide a better living environment, all at the same time! Comparing a good passive house to a conventional one is like comparing a modern, dependable lightweight bike to the ballon-tired one-speed bikes of the 1950s.

I Thought You Said It Was Simple?

It is. Every material and principle incorporated into passive solar design is common and in everyday use. The melting of an ice cube or the ability of a stone to stay warm long after sunset—these are the kind of considerations on which all passive design is based. The only trick is to learn the labels so that it is easier to understand and discuss. Then you can say "thermal mass" instead of having the say (each time you discuss the phenomenon) "the ability of a stone to stay warm long after sunset."

Air stratification
The tendency of heated air to rise and to arrange itself in layers with the warmest air at the top.

Conduction
The transfer of heat between objects by direct contact.

Evaporative cooling
Natural cooling caused by water's ability to absorb heat as it vaporizes.

Natural convection
The movement of heat through the movement of air or water.

Thermal radiation
The transfer of heat between objects by electromagnetic radiation.

British thermal unit (Btu)
A measure of energy. It is the amount of heat necessary to raise one pound of water one degree Fahrenheit. Heat loss and gain can be measured in Btus per hour and Btus per year.

Degree day
A unit used to measure the intensity of winter. The more degree days there are in total for the season, the cooler the climate.

Mean radiant temperature
The average temperature you experience from the combination of all of the various surface temperatures in a room—walls, floors, ceilings, furniture, and people.

R-Value
A measure of the insulating ability of any material or group of materials such as make up a wall or ceiling. The higher the R-value, the better the insulation and the slower the heat loss.

U-Value
A measure of the rate of heat loss through a wall or other part of a building. It is the reciprocal of the total R-values present. The lower the U-value, the slower the heat loss.

Phase-change storage materials
Meltable materials store heat when they change phase from solid to liquid form, and release that heat when they re-solidify. The heat released is called the "heat of fusion." These materials require less mass (and volume) to store the same amount of thermal energy than more conventional heat-storage materials. Only small changes in temperature are necessary to induce the phase change.

Thermal mass
Materials that store heat. Heavy, dense materials—concrete, stone, and even water—store a lot of heat in a small volume (compared with most lightweight materials) and release it when needed.

Glazing
Layers of glass or plastic, used in windows and other solar devices for admitting light and trapping heat.

Insulation
Materials that conduct heat poorly and thereby slow down heat loss from an object or space.

Movable insulation
Insulating curtains, shutters, and shades that cover windows and other glazing at night to reduce heat loss.

Reflectors
Shiny surfaces for bouncing sunlight or heat to where it's needed.

Windows
Windows let light (and heat) in (and out).

Shading
Measures for blocking out unwanted sunlight that can overheat the house.

What, More Definitions?

No, just a preview. The following are the passive
systems covered by this book. As with everything
else, there are both advantages and disadvantages.
For most of us, though, the advantages far outweigh
the disadvantages.

Solar rooms

Solar-heated rooms such as sunrooms, greenhouses,
sun porches and solariums are the hands-down
favorite passive solar system. They give a house extra
solar-heated living space; they provide a feeling of
spaciousness, a sense of outdoors; they act as buffer
zones between the house and outdoor weather
extremes. Solar rooms are often referred to as
"attached sunspaces."

Advantages. Solar rooms can greatly improve the
interior "climate" of a house. Although solar room
temperature swings may be large, since plants can
tolerate much wider swings that people can, house
temperature swings can still remain small (3–8°F).
Solar rooms can add humidity to the house air if
desired. A solar room can become additional living
space for a relatively low cost. They are readily
adaptable to present homes. And people love them!

Disadvantages. Improperly designed or built solar
rooms will not work well. Although construction
costs can be kept down, good-quality construction is
expensive. Factory-built kits can also be expensive.

Solar windows

When you make a conscious effort to place lots of glass on the south side of your house, feel free to call the extra glass "solar windows." The sunlight that enters your house directly through windows turns into heat. Some of the heat is used immediately. Floors, walls, ceilings, and furniture store the excess heat. Movable insulation can cover the windows at night to slow heat loss. South-facing vertical glass takes advantage of the winter sun's low position in the sky. In the summer, when the sun is high, the glass is easily shaded by roof overhang or trees. Solar windows are often referred to as "direct gain" systems.

Advantages. Everyone can use this simplest of all solar designs. In fact, most of us already do, but not nearly as much as we should. Solar windows are inexpensive and they provide a light and airy feeling.

Disadvantages. Not everyone appreciates sunshine pouring in all day. Many people enjoy the extra heat and the higher temperatures, but sunlight can fade fabrics, and too much glass may cause too much glare.

Solar chimneys

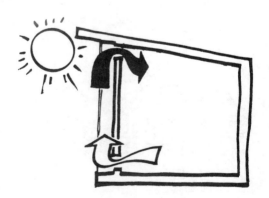

Air is warmed when it touches a solar-heated surface. The warmed air rises, and cooler air is drawn in to replace it. This is what happens in an ordinary chimney. The process of natural convection can occur in a continuous loop between your house and a solar collector attached to its south wall. As the air in the solar collector is heated, it expands, rises, and enters the house. Cooler house air is drawn into the collector to take its place. This is why these "solar chimneys" are usually referred to as "convective loops" or "thermosiphoning air panels."

Advantages. Solar chimneys are very simple and avoid many of the problems of direct gain systems, such as glare and heat loss. Also, they're easy to attach to existing homes and they're automatically insulated against heat loss at night.

Disadvantages. Like direct gain, too large a system may result in higher than normal temperatures in your house. Careful design and construction is required to ensure proper efficiency and durability.

Solar walls

When the mass for absorbing the sun's heat is located right inside the glass, you have a "solar wall." The wall heats up as the sunlight passes through the glass, strikes the surface, and changes to heat. Dark paint on the wall helps absorb more heat by reducing the amount of reflectance. Heat is then conducted through the wall and into the house.

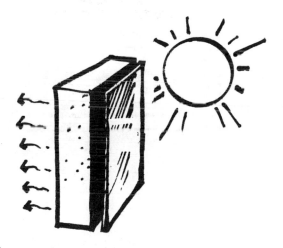

Another type of solar wall substitutes water for masonry. Tall cylinders of water, 55-gallon barrels, and specially-fabricated water walls are common. The water containers radiate their solar heat directly to the room. These wall are often referred to as "thermal storage walls" or "mass walls."

Advantages. Thermal storage walls have many of the same advantages as convective loops and simultaneously solve the heat storage problem. The mass is right where it belongs—in the sun. The thermal mass keeps the house at pretty even temperatures nearly 24 hours per day.

Disadvantages. The wall also loses heat back to the out-of-doors through the glass. Triple glazing or movable insulation solves this problem in cold climates but can be costly. Construction of the wall can be expensive and may, in some cases, reduce available floor area.

Solar roofs

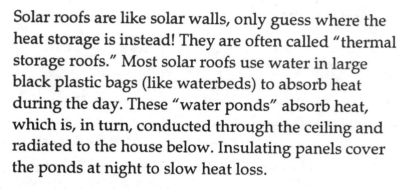
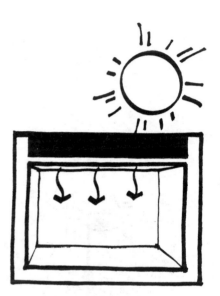

Solar roofs are like solar walls, only guess where the heat storage is instead! They are often called "thermal storage roofs." Most solar roofs use water in large black plastic bags (like waterbeds) to absorb heat during the day. These "water ponds" absorb heat, which is, in turn, conducted through the ceiling and radiated to the house below. Insulating panels cover the ponds at night to slow heat loss.

Solar roofs can, in certain climates, cool your house during the summer—the water absorbs heat from the house below during the day, and radiates it to the cool night sky. Insulating panels shade the roof ponds during the day.

Advantages. These systems can, in some climates, provide for all your heating and cooling needs. And they can do so while keeping you as comfortable, or even more comfortable, than almost any other type of heating system, whether solar or conventional.

Disadvantages. Solar ponds require careful design, engineering, and construction. Their efficiency and cost effectiveness are not nearly as good in cold, partly cloudy or cloudy climates as they are in dry, sunny southern ones.

Solar Building Design

Conservation First

To most people, solar energy is more glamorous than energy conservation. Basking in the winter sun can be pleasant, even though the temperature is below freezing. But you would never go out without "buttoning up" first. It should be just as obvious that it makes no sense to leave your house out in the cold without buttoning it up first.

In any climate where heating or cooling is a big thing, solar design done in combination with energy conservation works best. Conservation always pays off in savings faster than any other energy strategy.

The whole idea is simple. A tight, energy-conserving, passive solar home may reduce energy costs by 50 to 90 percent, depending on climate.

How heat is lost

Most heat is lost through either conduction or infiltration. Conduction losses occur as heat escapes through the roof and walls. Infiltration means losing heat through warm air escaping and being replaced by cold air drafts seeping into the house through cracks around doors, windows, and around the foundation, so plug them up—but be sure to arrange for proper ventilation as well.

13

Save cooling costs too!

In climates where energy is used for cooling, conservation comes first, too! People do not sit in the hot sun for hours without shielding themselves. So shading, ventilating, and insulating to keep heat out and cool air in are important to summer cooling. These simple, economic measures lower both the size and cost of air conditioning equipment, reduce cooling bills, and improve comfort by lessening the often large differences felt between indoor and outdoor temperatures and humidity levels when air conditioning is used.

In terms of costs, energy conservation measures fall into three categories: free, cheap and economical. You just can't miss. Let's talk about some specific energy conservation measures, starting with the free ones.

Free conservation

• Lowering thermostat settings to 68 degrees or lower during the day, and 55 degrees or lower at night.

• Reducing hot-water-tank temperatures to 120 degrees (or 140 degrees if dishwasher instructions require).

• Adding water-saving shower heads.

• Closing chimney dampers and blocking off unused fireplaces.

• Shutting off unused rooms.

• Turning off lights.

• Wearing sweaters.

• For cooling, get used to slightly warmer temperatures or turn the air conditioner off altogether and open windows.

When these measures are used in combination, 20 percent savings is easily obtainable at no extra cost.

Cheap conservation

- Maintaining the efficiency of your heating system through servicing check-ups.

- Caulking and weatherstripping windows and doors to seal infiltration cracks.

- Installing a clock thermostat for automatic set-backs.

- Adding sheets of plastic to windows that are the big heat losers.

- For cooling, shades, awnings—and trellises with plant growth—block out the sun.

- Fans are much less expensive to run than air conditioners.

Economical conservation

- Adding extra insulation in attics and walls and around foundations.

- Adding storm windows and doors, or replacing older windows with tight new ones.

- Covering windows with night insulation.

- Adding a vapor barrier wherever possible.

- Replacing an old, inefficient burner or furnace with a new one.

- Installing a woodburning furnace or stove.

- Adding an airtight entryway or planting a windbreak of trees. (But not blocking out sunlight from the south!)

These measures do require an initial dollar investment, but they almost always make economic sense, even if a bank loan is necessary to finance the investment.

So far, we have talked about energy conservation primarily in terms of retrofitting existing homes. Planning energy conservation into a new house design is a whole new game economically, one in which you stand to reduce energy bills even further. (Remember, energy conservation is the first step to efficient solar heat!) By planning before construction, many of the economical measures become cheap or free. Extra insulation and a vapor barrier, for example, are very inexpensive in new construction as compared to retrofitting with the same materials. The small extra cost of doing it right may be offset by lower costs of other items, such as a smaller furnace, as well as by energy savings.

Now we want to elaborate upon some conservation options. They are presented in their relative order of cost effectiveness.

Caulking and weatherstripping

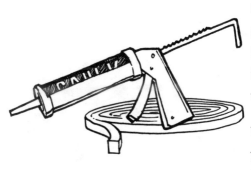

Caulking and weatherstripping reduce air infiltration by sealing cracks around windows, doors, wall outlets and the foundation. These materials are inexpensive and easy to use. Caulks include silicones, urethanes and materials with an oil or latex base.

Weatherstripping is made of felt, foam-backed wood, and vinyl or steel strips. A supplier can inform you of the types, costs, life expectancies and uses of these materials. One-to-five-year paybacks are usual, but sealing off big cracks may pay for itself in a matter of months.

Insulation

Insulating materials are assigned "R-values," a rating of how well each material resists the conduction of heat energy. The higher the R-value per inch thickness of material, the more effective the insulation.

Insulation types include fiberglass (which is available in a range of thicknesses), bags of loose cellulose, blow-in foams, and plastic foam sheets or boards. Check with reputable suppliers for the possible fire or health hazards of these materials as well as for their comparative durability. Installers can recommend types and amounts of insulation best suited to your particular house.

Heavy insulation in the roof, exterior walls and foundation reduces conduction losses. Recommended R-values of insulation for new construction in moderate climates are R-38 in attics and ceilings, R-19 in walls, R-19 in floors over crawl spaces, and R-11 around foundations. In severe climates, twice this much insulation should be used.

These standards should be followed or exceeded in all new construction, but they may not be as easy to reach in existing houses. Generally, in older houses the more insulation the better, and the investment will be a good one.

Vapor barriers

Vapor barriers protect wall and ceiling insulation from moisture. As warm house air seeps through walls and ceilings to the cold outside, it carries moisture with it into the insulation. If moisture condenses and is trapped, the insulation loses much of its effectiveness. In severe cases, moisture can cause wood to rot.

Adding vapor barriers to existing houses is difficult without tearing down interior walls. However, vapor-barrier paints and vinyl wallpapers are available for existing houses. Ridge and soffit vents for attic ventilation help carry away moisture—and help cool the attic in warm months. In humid climates, consult local builders, since vapor barriers are not used in some parts of the country where moisture moves from the outside into walls during the summer.

Storm windows and doors

A window with a single pane of glass (called single-glazed) can lose 10 times the heat of an R-11 wall (3.5 inches of fiberglass insulation) and 20 times the heat of an R-19 wall (6 inches of fiberglass insulation). Adding a second layer of glazing can save one gallon of oil each winter per square foot of window area in cold climates. Adding two layers (resulting in triple-glazing) can save nearly two gallons of oil per square foot of window area.

Storm windows are available with wood, aluminum or plastic sash. Aluminum insulates least, and wood insulates best. Storm windows come with single or double panes of glass. Placing clear plastic sheets over windows is the least expensive solution.

Thermal window shades

Thermal shades, shutters, or heavy curtains can reduce heat loss through windows at night by up to 80 percent. Many types of night insulation can be hand-made. Others can be purchased and professionally installed. A snug fit and tight edges all around are important for high effectiveness. Combining night insulation with double or triple glazing can be a great deterrent to night-time heat loss through windows.

Human comfort

Regardless of how fuel bills are reduced, the primary purpose of energy consumption for heating and cooling is to keep people comfortable. Passive solar design is a natural strategy for accomplishing this.

Our bodies use three basic mechanisms for maintaining comfort: convection, evaporation/respiration, and radiation. Air temperature, humidity, air speed, and mean radiant temperature all influence how we use our comfort control mechanisms.

Perhaps mean radiant temperature is least understood. Mean radiant temperature (mrt) refers to the average temperature we feel as a result of radiant energy emanating from all surfaces of a room: interior walls, windows, ceilings, floors, and furniture. It combines with the room air temperature to produce an overall comfort sensation, and different combinations of mean radiant temperature and room temperature can produce the same comfort sensation. For example, if the air temperature is 49°F and the mean radiant temperature is 85°F, you will feel as thought it is actually about 70°F. The same holds true for the combination of an air temperature of 84°F plus a mean radiant temperature of 60°F.

Many passive systems use warm surfaces to keep a house comfortable.

The higher mean radiant temperatures provide comfort at lower air temperatures. Most people prefer this comfort balance to the more common comfort balance found in conventional houses, where warm air is surrounded by cool or cold surfaces. In other words, because you are surrounded by warm surfaces, a passive house makes you "feel" warmer, even with room temperatures several degrees lower than you might have in a conventional house.

Once you've become accustomed to passive warmth, conventionally heated rooms feel cool and drafty, even at identical air temperatures. Interior surfaces of thickly insulated walls, floors, and roofs are warmer than those that are poorly insulated. The same holds true for multi-paned windows compared with single-glazed. Thus, energy conservation enhances mean radiant temperature and is a good companion to passive solar in providing comfort.

Because lower house temperatures result in less heat loss, even more energy can be saved than is normally calculated.

The following combinations of temperatures produce the common comfort sensation of 70°F. Notice that a combination of lower mean radiant temperature and higher air temperature can make you feel cooler in summer, too.

Mean radiant temperature:	85	80	75	70	65	60	55
Air temperature:	49	56	64	70	77	84	91

Shades and awnings

To stay cool during the summer, keep the sun out of the house during the day. Inexpensive interior shades from a hardware store are least effective but work well for the small investment, and can reduce winter heat loss, when closed, by almost 25 percent. Exterior awnings are becoming popular again and do a good job of shading. Deciduous trees provide ideal summer shading and shed their leaves to allow winter sun in. Summer shading by whatever means is a natural first step in reducing cooling costs.

Taking steps

Take time to get sound advice and to implement energy conservation measures properly. Although procedures are often simple, inadequate or faulty installations can reduce insulating effectiveness, or even damage the house. But, without a doubt, energy conservation will in fuel savings pay you well for your effort.

Solar Retrofit

Solar retrofitting, or adding solar features to existing homes, is one of the most exciting challenges in the field. There are a lot of existing homes and all of them use energy, often more than necessary. For many reasons—economic, environmental, historic, aesthetic, or purely sentimental—we don't want to just discard older homes or other valuable buildings. Solar renovating or retrofitting is a viable option to consider. Take time to compare options before choosing. Some solar retrofits will suit your existing structure better than others.

Go south

Research for your solar retrofit starts very simply. Find south. Every home has a south-facing wall, or corner, at worst. If the compass does not yield a perfect long wall facing due south, don't give up. Orientation can be up to 30 degrees off either to the east or west of south and still be effective for solar collection.

If when looking southerly, you don't find a high rise, you're in luck. An obstruction casting a shadow on the house in winter reduces solar collection unless it can be removed or, like a deciduous tree, looses its leaves when you will need the winter sun. Summer shading from deciduous trees is a cooling advantage, too. Solar orientation and shading factors are only the first steps in evaluating site suitability.

If the site checks out so far, think of the passive solar options described in Chapter 1. Which ones make sense for you? That will depend upon the design of your home, its position on your lot, and the dollar investment you are prepared to make. For example, a solar room will make sense if you like gardening or want extra living space in addition to solar heat.

If you want solar heat, and privacy is needed because your south side faces a street, a solar chimney for a frame house or a solar wall for a masonry or frame house makes the most sense. Adding more south-facing windows is a simple, efficient solution in many cases. If the only available south-facing surface is a roof, a solar attic may be a natural choice.

A solar retrofit, like any remodeling work, will cost money, but compared to what? If you convert the home you have to solar instead of building a new one, you could save a lot of money by comparison and substantially reduce heating and cooling costs as well.

Passive solar retrofits come in many sizes as well as costs. A greenhouse solar room, for example, can vary from light lean-to framing covered with plastic glazing to custom built, with triple-glazed windows, built-in shutters and thermal-mass heat storage. Both are appropriate for some uses and both function well. The cost is the variable, so you must decide what you want to spend.

One more bit of information you need is energy flow in and out of buildings. Heat and light flow through windows. Heat travels through walls also. The important point is that there are two primary flows of heat in and out of the house. One is solar radiation inward; the other is heat escaping from your house in cold weather and seeping in during hot weather. Both vary considerably depending on the time of day and the season of the year. The amount of that heat loss or gain depends greatly on your location.

Add glazing

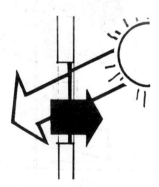

Adding transparent glazing to a house is the basic strategy in solar retrofitting. It lets solar heat in, and traps it to prevent heat loss out. Solar retrofit strategy assumes that conservation measures have been taken first.

A. Double and triple window glazing. A single layer of glass allows high levels of both solar radiation in and heat loss out. When the sun is shining, the house heats up quickly. Yet, when the temperature drops, heat loss increases quickly. When a second layer of glazing is added to an existing window, it greatly reduces heat loss, but reduces solar heat gain only slightly: it reduces solar gain by about 18%, but reduces heat loss by about 50%. A third layer of glass reduces solar heat gain by another 18%, but heat loss is reduced by an additional one-third. Therefore, a second, and even a third, layer of glazing is often cost effective. Low-emissivity glazings and special gas fillings can significantly reduce heat loss over their clear counterparts, at little or no additional cost.

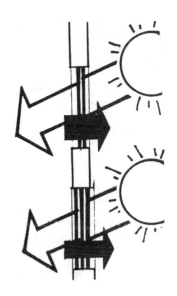

B. Solar chimneys. The same principles apply if you convert south-facing walls into windows, perhaps the simplest solar retrofit. A poorly-insulated wall allows small amounts of solar heat gain. A well-insulated wall allows little or no solar gain. Whereas the poorly insulated wall has huge heat losses, the well-insulated wall loses very little. When the outdoor temperature drops, heat loss through the walls increases quickly, but not as quickly as it does through windows. When glazing is added next to the prepared wall surface of a house, it transforms the wall into a solar chimney collector. This increases considerably the amount of solar energy coming into the house.

Such collectors take a short time before they heat up and start producing heat, so they do not provide energy quite as quickly as windows do. Nor do they provide as much heat as windows do. However, the heat loss from the house through the walls is substantially less than through the windows (unless, of course, the windows are covered with insulation at night.) The net result is that more energy is gained through solar chimneys than is lost. And, if properly constructed, solar chimneys can produce more energy than windows can.

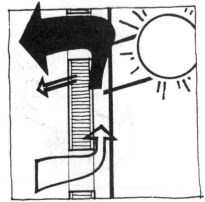

C. Solar chimneys and thermal mass. Adding a layer of glazing next to an uninsulated masonry wall significantly reduces heat loss and, in fact, produces considerable solar heat gain. Brick, stone, adobe, and concrete walls have high rates of heat loss, even if they are thick. If they face east, west, or north, insulate them, preferably with the insulation on the outside. But if they face south, cover them with sheets of glass or durable plastic and capture the sun's rays. Much of the sun's heat is absorbed by the wall, delaying the time when the house receives the heat. Also, because the sun-warmed wall loses heat back to the outdoors, the net energy gain is not as great as it is through windows. But the solar heat enters the house slowly, over a long period of time, making overheating much less of a problem than with solar windows, and keeping the house warm well into the night.

D. Solar rooms. Vertical glazing offers only a small, dead air space over an exterior wall surface. If the glazing is installed instead in a lean-to fashion, the air space becomes large and can be called a solar room. This "sunspace," in a sense, absorbs the shock of outdoor weather extremes, tempering their effect on the house while also providing solar heat. The heat loss from the house is no longer to the outdoors, but rather to this large air space, which is nearly always warmer than the outdoors. This makes the rate and amount of heat loss from the house much lower.

If the wall of the building is wood framed, a solar room is likely to experience wide temperature fluctuations. If the wall is of solid masonry, then the fluctuations will be much smaller. The thermal mass of masonry or earthen floors reduces temperature fluctuations, too.

Natural daylighting

Do not underestimate the bonus of natural daylighting, which passive solar designs can provide. In some big buildings, solar glazing may save more energy and money by reducing electric light bills than it saves by reducing fuel bills, and lighting engineers feel that properly located lighting from sidewalls can be two to three times as effective as artificial overhead lighting. For houses, the extra light from solar windows and solar rooms can add immeasurable pleasure and a living experience far surpassing any you've had before.

Solar Position

We all realize that the sun doesn't stay in one part of the sky all day, and that its path varies from season to season and from state to state. Fortunately, its movements are completely predictable, widely published, and easy to understand. No guesswork is involved: from the sun chart for your latitude (see Appendix 1) you can find quite easily where the sun will be at any hour, at any season, and from that information you can see how and where solar installations (and summer sunshades) must be placed in order to respond to the sun where you live.

Here's a nice surprise: the quantity of solar energy that penetrates a south-facing window on an average sunny day in the *winter* is greater than that through the same window on an average sunny day in the *summer!* Here are the reasons:

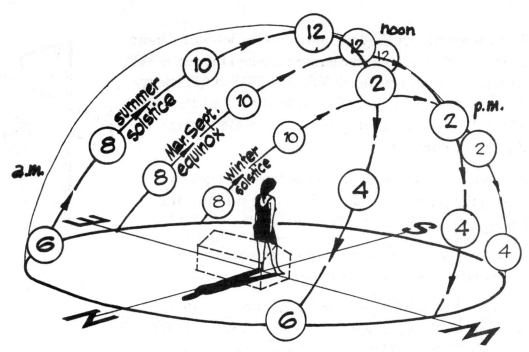

Instructions: Stand at the line of your proposed solar wall, face south, look at the numbered suns in the sky, wait there for 12 months until all 17 have appeared, then continue reading this book.

1. Although there are more daylight hours during the summer, there are more possible hours for sunshine to strike a south-facing window in winter. If you live at 35 degrees north latitude, for example, there are fourteen hours of sunshine on June 21. But at that position the sun remains north of east until after 8:30 a.m. and moves to north of west before 3:30 p.m., so that direct sunshine occurs for only seven hours on the south-facing wall. On December 21, however, the sun shines on the south wall for the full ten hours of daylight.

2. The intensity of sunlight is approximately the same in summer as in winter. The slightly shorter distance between the earth and the sun during the winter than during the summer is offset by the extra distance that the rays must travel through the atmosphere in the winter when the sun is low in the sky.

3. In the winter, the lower sun strikes the windows more nearly head-on than in the summer, when the sun is higher. At 35 degrees north latitude, 170 Btu's of energy may strike a square foot of window during an average winter hour, whereas only 100 strike on the average during the summer.

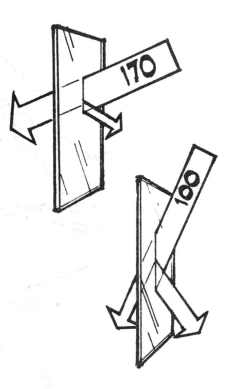

4. In winter, more sunlight passes through glass by hitting the window head-on. But in the summer, the high-angle rays tend to reflect off the glass.

5. With proper shading, windows can be shielded from most of the direct summer radiation.

About twice as much solar radiation is transmitted through south-facing windows in the winter as in the summer. If the windows are summer-shaded, the difference is even greater.

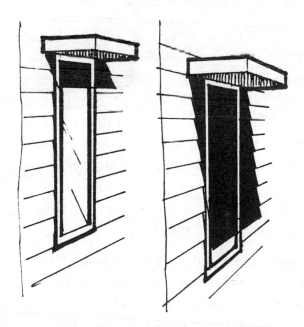

In passive systems, tilted surfaces such as roofs are used less often than vertical surfaces. With reflective surfaces such as snow on the ground, a south-facing vertical surface actually receives more energy during the middle of the winter than a south-facing tilted one. Therefore, during the primary heating months there is little advantage to using tilted rather than vertical, south-facing surfaces. In fact, for more northern latitudes, the difference is insignificant.

Tilted glazing, whether in collectors or skylights, tends to be more costly to build and more prone to leaks. It is also harder to shade and, if left unshaded, can more easily overheat the house in the summer than vertical glazing. Roofs are less likely than walls to be shaded by trees or buildings during the winter, and they have large surfaces for collecting solar energy. Unfortunately, it is difficult to cover them with insulation to reduce heat loss at night.

The Site

If your site does not have proper solar exposure because it slopes sharply north or is shaded darkly by evergreens or large buildings, you have little opportunity to add solar heating features to a house designed for the site. Here's what to look for.

Lot orientation

South-facing houses assure lower energy consumption, during both summer and winter. This does not mean that houses have to face rigidly southward. A designer who understands passive solar principles can devise dozens of practical solutions. The site or lot itself does not have to face south, as long as the building itself is oriented southward.

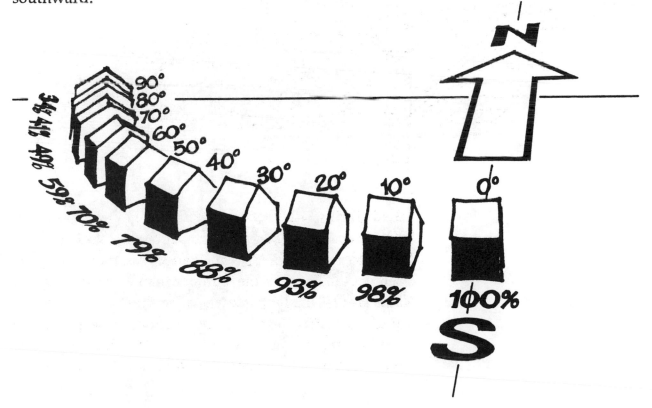

A lot that slopes sharply north is, of course, very difficult to work with, and south-sloping lots are preferred. Once land developers understand passive principles, they can plan for solar subdivisions, with the cost approximately the same as for conventional subdivisions.

Setback flexibility and minimum lot size

Deep house lots which have narrow street frontage reduce the surface area of summer heat-producing asphalt streets. Higher housing densities can reduce travel distances and times and subsequent energy use. Flexible zoning laws can permit houses to be located near the edges of their lots, thus minimizing the potential of shading from adjacent neighbors. Long-term, shade-free rights to the sun are necessary to guarantee adequate sunlight for the life of the house.

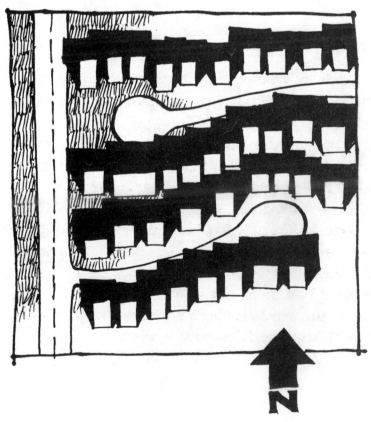

Landscaping

Proper landscaping can offer beauty as well as comfort
and energy savings, both in winter and in summer.
Evergreens can greatly slow arctic winds. For most of the
country, these winds come from the west, north, and
northwest. Large deciduous trees appropriate to your
region can provide shade and summer cooling. They are
most effective on the east, west, and south sides of the
house. Most, but not all, deciduous trees shed their leaves
in the winter to let the warm sun in. Glaring, unshaded
asphalt creates desert-like conditions, placing a higher air
conditioning load on buildings. Well-shaded and
landscaped paving that is porous to rain and does not
absorb heat has a much less severe effect.

Street widths

Narrow streets save valuable land and can be shaded more easily than wide ones. They are more pleasant than wide multi-laned streets and are safer for bicyclists, pedestrians, and motorists. They reduce the heat load on people using them, and they also reduce traffic speeds. Parking bays, rather than on-street parking, can promote shading both over the bays and over the narrower streets. Pedestrian walks and bicycle paths are far more readily integrated into such plans.

Length/width ratios

In the northern part of the country, south sides of houses receive nearly twice as much radiation in the winter as in the summer. This is because the sun is lower in the sky during the winter. In the summer, the sun is high in the sky, and the sun does not shine directly on south walls for a very long period of time. Houses in the south gain even more on south sides in the winter than in the summer. East and west walls receive two and a half times more sunshine in the summer than in the winter.

Therefore, the best houses are longer in the east/west direction, and the poorest are longer in the north/south direction. A square house is neither the best nor the worst. Remember, however, that a square building is often the most efficient in terms of layout, economy of materials, and lower heat loss per square foot of floor area.

A house poorly shaped for best solar gain can be improved by covering the south wall with windows and other passive systems and minimizing the number and size of windows facing other directions.

Solar Rooms

Solar rooms are spaces with large areas of south-facing windows. Examples include sun rooms, solariums, sun porches and greenhouses. Without ventilation or thermal mass, the temperatures of such spaces will fluctuate widely. Temperatures of conventional greenhouses, for example, can rise to over 100 degrees on sunny winter days and then drop to below freezing at night. Almost always, however, a solar room is warmer than the outdoor temperature, thus reducing heat loss from the building where the room is attached.

Which Direction?

Solar rooms that face east or west do not work as well for heating as those that face south. They supply less heat during the winter and may provide much too much in summer. However, an east-facing greenhouse can give morning light, which plants like; it can also be a buffer zone to reduce heat loss from the house throughout the day.

If an east-facing solar room seems to be a good solution to either site or building problems, locate spaces such as kitchens on the east side of the house, next to or behind the solar room, to take advantage of the morning light and heat. Then living rooms and bedrooms located on the west side, which usually remain cool during the day, will become warm in the afternoon from the heat gained from the west.

33

A leakproof skylight detail

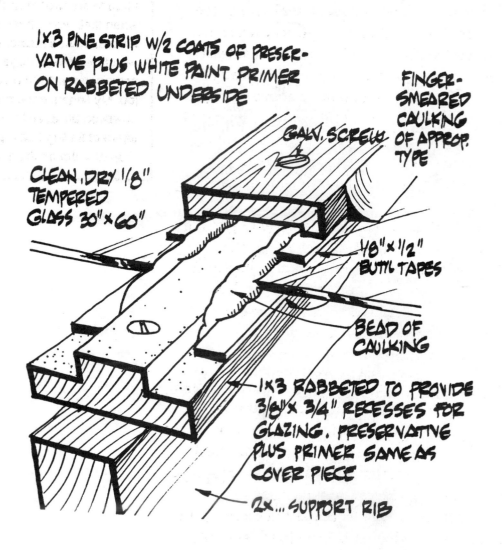

1×3 PINE STRIP W/2 COATS OF PRESERVATIVE PLUS WHITE PAINT PRIMER ON RABBETED UNDERSIDE

FINGER-SMEARED CAULKING OF APPROP. TYPE

GALV. SCREW

CLEAN, DRY 1/8" TEMPERED GLASS 30" × 60"

1/8" × 1/2" BUTYL TAPES

BEAD OF CAULKING

1×3 RABBETED TO PROVIDE 3/8" × 3/4" RECESSES FOR GLAZING. PRESERVATIVE PLUS PRIMER SAME AS COVER PIECE

2×...SUPPORT RIB

Glazing

A solar room is designed both to maximize solar gain and to minimize heat loss. Usually, only the south-facing walls and roof are glazed, while the east and west walls are well insulated. If at least two layers of glass or plastic are used instead of one, this type of room will remain above freezing most of the winter in all but the coldest climates of this country. However, for maximum heat savings, three and even four layers of glass and plastic should be used where winters have more than 5000 degree days.

Keep in mind that each additional layer of glazing blocks additional sunlight. Therefore, for the highest possible light transmission, the third and fourth layers must be a very clear film, such as Teflon™ or Tedlar™. Each layer must be sealed tightly to prevent structural damage from possible moisture condensation between glazings.

Glazing for solar rooms should be vertical or sloped no more than 30 degrees from vertical (at least 60 degrees from horizontal). Before you build, however, talk to everyone you can find who has ever used glass in a sloping position and ask about leaks. If you can find someone who can convince you of a leakproof system, do not let any detail escape your attention. Also, read the fine print in the sealants literature. Some silicones are attacked by mildew, many won't stick to wood, and all must be applied only to super clean surfaces.

A Glazing Experience

I had to fire two "expert" glazers when their work on my skylight leaked. Then a master carpenter reinstalled the 92 large panes and actually got a 100-percent leakproof job. My fingers are crossed, however, because I'm always half afraid of finding that first tell-tale drop of water on the floor.

A solar room results when the space between the glazing and the wall is greatly enlarged. This space, in a sense, absorbs the shock of outdoor weather extremes, tempering their effect on the house while also providing solar heat.

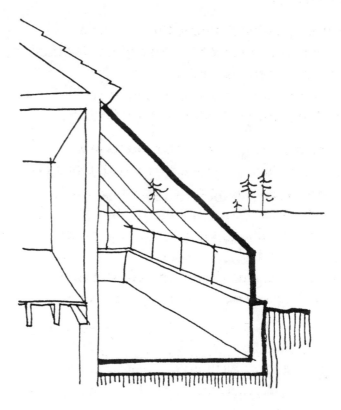

Insulation

For maximum sunshine, and for minimum heat loss at night, movable insulation is used in combination with glazing. This can be tough to do, however. Some of the tricky design and construction problems include storing the insulation out of the way during the day, interfering with furniture or plants while moving the insulation, and obtaining tight seals against the glazing when the insulation is closed. Additional considerations include the need for insulation to resist mold, plant and insect life, and moisture damage.

Heat Storage

As with other passive systems, thermal mass enhances the performance of a solar room. Thermal storage mass moderates temperature swings, provides more stable growing temperatures for plants, and increases overall heating efficiency.

Many of the most successful solar rooms are separated from the house by a heavy wall that stores the heat. The wall, built of concrete, stone, brick, or adobe, conducts heat (slowly) into the house. At the same time, the wall keeps the solar room cooler during the day and warmer at night. To determine what you need, use the design and construction information for solar walls, without the glazing.

Earth, concrete, or the floors store considerable heat. So do foundation walls if insulated on the outside. Be sure to use insulation with an R-value of at least R-12 (3 inches of polystyrene). Insulate at least 3 to 4 feet deep and more in deep-frost country. This gives better protection than insulating 2 feet or so horizontally under the floor.

Heat-storing capability also can be obtained with 55-gallon drums, plastic jugs, or other containers filled with water. Two to four gallons of water per square foot of glazing is probably adequate for most solar rooms.

When solar rooms larger than 200 square feet reach 90 degrees, a fan can be used to circulate the collected heat. The hot air can be blown horizontally through a 2-foot deep bed of stones below the solar room floor.

Stone beds can also be built beneath the floor of the house and should not be insulated from it. Then the heat will rise naturally through the stone beds and into the house.

Two cubic feet of ordinary washed stone per square foot of glazing is sufficient. Use a fan capable of moving about 10 cubic feet of air per minute for each square foot of glazing. Potato-sized stones, larger than the usual 3⁄4 inch to 1 1⁄2 inch size, will allow freer air movement. Consult with a local mechanical engineer or heating contractor for the best fan and ducting design. (Keep it simple!)

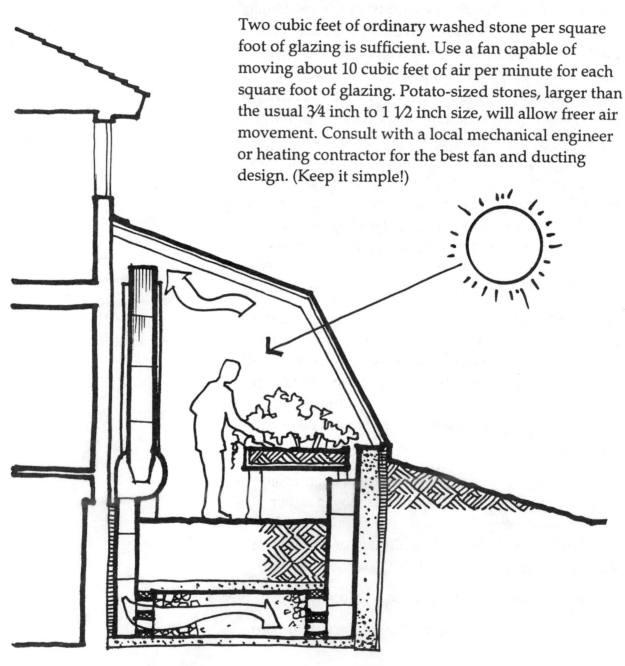

Costs

Solar rooms can be relatively simple to build, yet they can be very expensive if they are of the same quality and durability as the rest of your house. For example, with a few hundred dollars worth of materials you can build a simple, wood-framed addition to your house to support thin-film plastics. The resulting enclosure will provide considerable heat, especially if it is not used for growing plants. On the other hand, good craftsmanship and quality materials can result in costs of several thousand dollars. In general, solar rooms are most economical when you use them for more than providing heat, and when they are built to a quality that will both enhance the value of your house and appreciate in value as your house does.

Large Solar Rooms

Most of the information in this chapter is applicable for relatively small solar rooms of 100 to 200 square feet. Unless your house is superinsulated or in a mild climate, a solar room of this size will provide less than 25 percent of your heat. For big leaky houses, small solar rooms will provide as little as 5 or 10 percent of the heat.

Another way of approaching the use of solar rooms to help heat your house is to think of them as rather large spaces incorporated into, rather than attached onto, your house. There are a number of advantages to this approach:

• Both the solar room and house will lose less heat.

• Heat will move easily from the solar room to your house.

- Natural light can be made to penetrate deep into your house.

- The solar room can be easily heated by the house if necessary and so is unlikely to freeze.

- The solar room can be readily used as an expanded living space.

- You can build your house compactly and the solar room will provide a feeling of large exterior wall and window area.

- The costs can be less than for solar rooms that are simply added onto conventional house designs.

- The excess humidity of the solar room can be somewhat reduced by, and profitably used by, the excessively dry winter house.

Growing plants—some things to remember

Some people like to grow plants in a solar room. Warm soil and sufficient light are critical for successful plant growth. Remember that the multiple layers of glass or plastic you may need to use will reduce light levels, a crucial issue in climates with below-average sunshine. Circulation of warm air through gravel beds under the soil can raise planting bed temperatures, increasing the growth rate of most plants.

Cold-weather plants can tolerate cold temperatures, sometimes even mildly freezing ones. Few house plants can be permitted to freeze, but many can endure rather cool temperatures. On the other hand, when warm, stable temperatures are required, the solar room must retain most of its solar heat; little heat should be allowed to move into the house. Three or four layers of glass or plastic (or movable insulation) and plenty of thermal mass are required to trap and contain the heat in cold climates.

Evaporation of water from planting beds, and transpiration by the plants, causes humidity. Each gallon of water thus vaporized uses roughly 8,000 Btu's, nearly the same amount of energy supplied by 5 square feet of glass on a sunny day. Also, water vaporization reduces peak temperatures. It may be undesirable to circulate moisture-laden air into the house, unless the house is very dry.

Attic Solar Rooms

Attics are often great places for solar rooms, particularly if their only purpose is to heat your home. Frame the roof in a conventional manner. Glaze the south slope with one sheet of glass or plastic. Insulate the end walls, the north roof, and the floor. Paint all of the surfaces black. When your house needs heat and the solar room is hot, a fan can circulate solar heated air from the attic to the house.

Be sure to insulate the sun-trap from the rest of the house. Place back-draft dampers on the air ducts to prevent house air from rising up into the attic at night when the attic is cold and the house is warm. This solar room design gets very cold on winter nights but heats up quickly when the sun shines.

Even though it has no thermal mass to store the heat that the house doesn't need, a solar attic can reduce fuel bills by up to 25 percent. In order to reduce fuel bills further, the design must be altered in a number of ways.

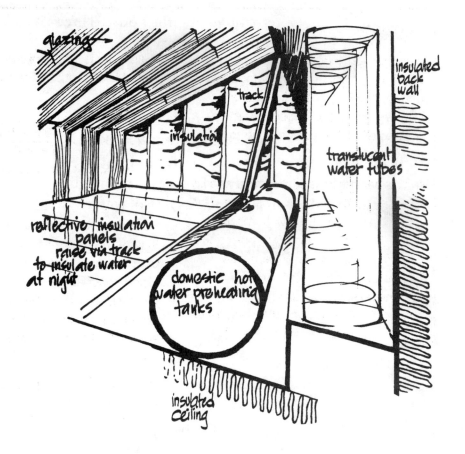

- The glazed portion of the roof must exceed 15 to 20 percent of the floor area of the house.

- Thermal mass must be added to the attic. This can be done by placing containers of water along the north wall of the attic.

- Movable insulation must cover the glazing at night to significantly reduce heat loss from the attic so as to trap and store the sun's heat. In climates of 3,000 degree days or less, double or triple glazing is an alternative to movable insulation.

Ventilation

Even the best designed solar rooms will require ventilation at times of too much heat or humidity or too little carbon dioxide. Ventilation should be able to replace all of the room's air up to six times each hour.

Natural ventilation is preferred to energy-consuming mechanical ventilation. The greatest amount of ventilation occurs when the exhaust vents are positioned as high as possible and the intake vents as low as possible. Air flow rates, and in turn, necessary vent sizes, can be estimated. The velocity of the air, in feet per minute, is approximately:

$$V = 486 \sqrt{\frac{h(T_0 - T_i)}{T_i + 460°}},$$

where

h = the height between the intake vent and

    exhaust vent

  = the temperature at the outlet vent, and

  = the temperature at the intake vent.

For example, if the outdoor temperature at the intake vent is 85 degrees, the temperature at the outlet vent is 100 degrees, and the height is 10 feet, then the velocity is:

$$V = 486 \sqrt{\frac{10(100 - 85)}{85 + 460}}$$

= 255 feet per minute (255 f/min).

For solar rooms that taper at the top (as in lean-to's), smaller values should be used. Air can carry 0.018 Btus per cubic foot for each degree it increases in temperature (0.018 Btu/cf-°F). The amount of heat exhausted through one square foot of vent each hour is, therefore:

(255 f/min) x 100° - 85°/sf)

x(0.018 Btu/cf- °F) x (60 min) = 4131 Btus

A representative value for solar heat gain through glass is 200 Btus per square foot each hour. Therefore, each square foot of vent can accommodate 20 square feet of glass.

Solar rooms must sometimes be ventilated in midwinter, when heat from the bright sunlight builds up too quickly to be dissipated through the house or absorbed by thermal mass. Just be sure that such vents are sealed tightly on winter nights and on cold sunless days.

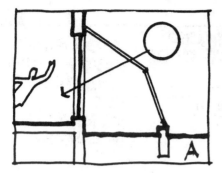

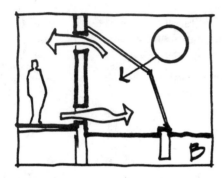

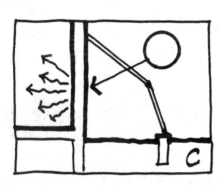

How to Get the Heat from the Solar Room Into your House

A. Windows

Windows in the walls between the solar room and the house let light pass right into the house (just as in direct gain systems), especially during winter months when the sun is low in the sky. The roof of the solar room can shade the windows during the summer, helping to keep the house cool. Since the house windows are protected from the weather, you can keep their construction simple and inexpensive.

B. Natural Air Movement

When your solar room is warm, just open the windows and doors and let the heat flow into the house. The higher the windows or other openings, the more heat will flow inside. You can use curtains to control the flow of heat. However, don't forget about odors, insects, and humidity. Screens over the openings are usually a must. A fan on a simple thermostat can regulate the amount of airflow into the house. The small extra expense will ensure that your "solar system" works when you're not home.

C. Conduction

Conduction through an unglazed thermal storage wall is one of the best ways of transferring heat into your house. The wall should not be insulated unless it is wood-framed. In the summer, the wall will protect the house from the solar room's heat. Be sure to protect the wood from the moisture of the room.

D. Gravel Beds

Use fans to blow warm air from the solar room through gravel beds under uninsulated floors of your house. Heat will radiate up through the floor into the room that is to be heated. The fans can be kept off during mild weather. Do not use fans to circulate the air from the gravel beds directly into the house, as there could be dampness and musty-smell problems. Radiant heat through the floor is much more effective and comfortable.

An All-Purpose Solar Room Design

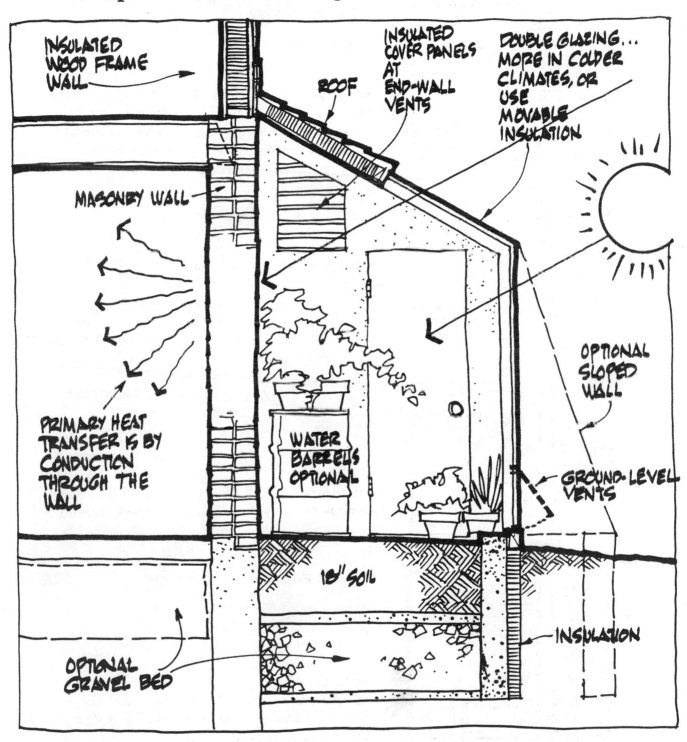

INSULATED WOOD FRAME WALL

INSULATED COVER PANELS AT END-WALL VENTS

DOUBLE GLAZING... MORE IN COLDER CLIMATES, OR USE MOVABLE INSULATION

ROOF

MASONRY WALL

PRIMARY HEAT TRANSFER IS BY CONDUCTION THROUGH THE WALL

WATER BARRELS OPTIONAL

OPTIONAL SLOPED WALL

GROUND-LEVEL VENTS

18" SOIL

INSULATION

OPTIONAL GRAVEL BED

It is difficult to sort through the confusing multitude of designs for solar rooms and to choose the one that makes the most sense. The all-purpose solar room shown on the preceding page will work throughout most of the country. Its net energy contribution to a house will vary depending on the severity of the climate.Summer temperatures can be kept close to outdoor temperatures with adequate ventilation. Mechanical ventilation and/or additional shading may be needed in hot, humid climates.

Winter temperatures in the solar room

• Up to 8,000 degree days and more than 70% possible sunshine: 45–85 °F.

• Up to 8,000 degree days and less than 70% possible sunshine: 35–85 °F with occasional need for backup heat.

• More than 8,000 degree days and more than 70 percent possible sunshine: 35–85 °F with occasional need for backup heat.

• More than 8,000 degree days and less than 70% possible sunshine: up to 85 °F with frequent need for backup heat.

Solar Windows

Sunlight falling gently through windows is by far the most common way for solar energy to heat a building. But loosely fitting, single-glazed windows usually lose more heat than they contribute. On the other hand, a properly-designed south window, with the addition of a reflective surface on the ground (such as snow, a pond, or an aluminized mirror), and with special glazing or an insulating cover at night, can supply up to twice as much solar energy to a building as a good "active" solar collector of the same surface area.

In one sense, of course, most of the world's buildings are already principally solar heated. Think about the huge, year-round heating job the sun does, using land and sea as thermal mass, to keep the whole world between −50 °F and +100 °F when, without the sun, we'd be at −473 °F all the time. That's *most of all the heating done on Earth*, and we'd do well to remember it when we hear talk of alternatives to solar energy use, or that solar can't really do the job.

Proper Design Criteria

The right timing

Sunlight must enter a house at only the right time of the year and the right times of the day. This simply means careful design in response to known solar geometry and climate.

The right amount of solar heat

If you desire fairly stable indoor temperatures, this must be engineered. If you desire a fairly wide range of temperatures, great—but don't just assume that you do, or that other occupants or your friends will find it comfortable, and then have to later excuse a poorly engineered system when it gets too hot. Too much glass and too little mass to absorb that heat is a common but unnecessary error in properly insulated passive solar houses.

The right type of glazing

Clear glass is both attractive and efficient and has its place. Special coatings on glass, such as low-emissivity (low-e) or reflective coatings, can reduce unwanted solar heat gains—as can other clear and translucent materials, such as diffusing glass, plastic films, fiberglass-based glazings and acrylics.

Minimum heat loss

Keep heat losses back out through the glazing as low as is practical. Use several layers of glazing according to the material, coating and climate. Cold climates also warrant movable insulation at night.

Control of glare and fading

Some people simply do not like working in direct sunlight. In fact, many people prefer the softer north light. Southern exposure means low fuel bills, but it also means window glare and squinting. Too much glass can also mean loss of privacy. Overhead light (such as from a skylight) is often a good compromise, offering solar gain with the least glare. In colder climates, however, this can mean added heat loss at night. Make sure the overhead glass is shaded during the summer!

Windows and Solar Collectors Compared

Properly designed solar collectors supply between 50,000 and 85,000 Btus per square foot of surface area per heating season in a climate where the sun shines half the time. (This is equivalent to he energy from one half to one gallon of home heating oil, or from 15 to 25 kwh or electricity, or from 100 to 140 cubic feet of gas.) Solar gain through a square foot of south-facing, double glass in the same climate is about 140,000 Btus. Conduction heat loss through that square foot (ignoring air infiltration for the moment) is about 70,000 Btus in a 5,000 degree day climate. The net contribution to the building, then, is 70,000 Btus (140,000 solar gain less 70,000 heat loss).

Therefore, in a climate like that of St. Louis, ordinary double-glazed south-facing windows can produce about the same amount of heat per square foot as solar collectors. Reflectors will boost heat production of both designs. Movable insulation, low-e glass (with or without argon gas), or triple-glazing can dramatically reduce heat loss from windows, greatly boosting their net energy input to the house.

Thermal Mass

The sun does not shine twenty-four hours a day, and thus, unlike a furnace, it is not waiting on call to supply us with heat whenever we need it. Therefore, when we depend on the sun for heat, we must do as nature does—store the sun's energy a number of ways. Plants use photosynthesis during the day, and then they rest at night. Lakes become heated during the day and maintain relatively constant temperatures day and night. For hundreds of years people have been growing and harvesting food during the summer and storing it for use during the winter. Indians of the American Southwest have for centuries used thick adobe walls that act like big thermal sponges to soak up large amounts of sunlight. As their exterior surfaces warm up during the day, the heat slowly moves throughout the adobe, protecting the interior from overheating.

Solar Gain

Appendix 2 provides month-by-month, hour-by-hour clear-day sunlight (or "insolation") data for vertical south-facing surfaces for six different northern latitudes. Together with Appendix 3 (U.S. sunshine maps for each month), it can be used to determine the approximate amount of solar radiation likely to come through south-facing glass anywhere in the United States at any time.

For example, from Appendix 2, the total clear-day solar radiation on a south-facing vertical surface in January at 40 degrees north latitude (Philadelphia, Kansas City) is 1726 Btus per square foot. In Kansas City, the "mean (average) percentage of possible sunshine" in January is 50 percent. (See the U.S. sunshine map for January.) Approximately 82 percent of the sunshine that hits a layer of ordinary glass during the day actually gets through it. Therefore, the average total amount of solar radiation penetrating one layer of vertical, south-facing, double-strength glass in Kansas City during the month of January is approximately

(31 days per month)

x (1726 Btus per square foot per day)
x (50 percent possible sunshine)
x (82 percent transmittance)

= 22,000 Btus per square foot per month.

Over the course of a normal heating season in a 50-percent-possible-sunshine climate, the total solar gain will be between 130,000 and 190,000 Btus per square foot. (A more accurate number may be obtained by doing the calculations on a month-by-month basis for each month of the heating season for a particular location, taking into account the heating needs of the particular house.)

If a second layer of clear glass is added to the first, about 82 percent of the light that penetrates the first layer will penetrate the second. Converting the first example, then, 18,000 Btus (0.82 x 22,000) are transmitted by double glass compared with 22,000 by single glass. But remember, heat losses are reduced by 50 percent when you do this!

These monthly solar gains can be roughly compared with the monthly heating demands of the house to determine the percent of the heat supplied by the sun. When solar provides less than 40 percent of the heat, the above analysis is relatively accurate for preliminary design purposes. However, a more detailed and rigorous analysis is required when the solar windows are large enough to be supplying more than 40 percent of the heating load.

In cold climates, 300 square feet of direct gain will supply roughly half the heat for a well-insulated, 1500 square foot house. Half as much area is needed in a mild climate.

At night, the walls cool off, allowing the adobe to soak up heat again the next day, thus keeping the houses cool.

In contrast, massive central masonry chimneys of New England colonial houses absorb any excess interior daytime warmth. The stored heat helps keep the houses warm well through the night.

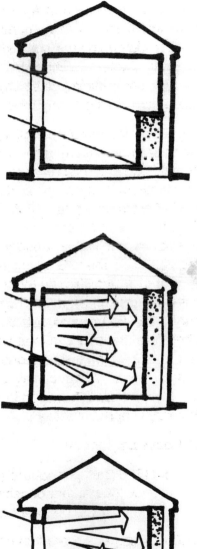

When massive materials are located inside houses where the sun can strike them directly, they combine the benefits of Southwest adobe and New England chimneys. The resulting "thermal mass" tempers the overheating effects of sunlight from large windows and absorbs excess energy for later use.

Generally, the more thermal mass the better. But how thick the mass should be depends on its location. If it's *too* thick, and is located between the sun and the space, the heat may not get through to the space before temperatures drop outside and the heat flow reverses direction. The more directly the sun strikes the mass, the less the house temperature will fluctuate. Unfortunately, thermal mass, such as brick walls, concrete floors, or water storage tubes, are often expensive and/or unsuitable to the homeowner. Thus, moderately sized solar windows, which require limited amounts of thermal mass, are often the best solution. Solar walls and/or solar rooms can supplement the solar windows to achieve the lowest possible fuel bills and the highest possible levels of comfort.

An unheated, lightweight house, such as a wood-framed one, drops in temperature relatively quickly, even if it is well-insulated. A heavy, massive, well-insulated structure built of concrete, brick, or stone maintains its temperature longer. To be most effective, the heavy materials should be on the inside of the insulating envelope of the house. When left unheated, a house that is well, and also buried into me side of a hill, cools off very slowly and eventually reaches a temperature close to that of the soil. Although earth is a good means of sheltering your house from the extremes of weather, soil—particularly wet soil—is a poor insulator and will draw heat out of the building endlessly if you don't insulate well.

If you prefer to close draperies to keep the sun out, or if you insist on wall-to-wall carpeting or big rugs, solar windows might not make sense for you. Alternatively, reconsider how you desire to furnish your house. The warmth of brick floors, walls, and, and the sensation of light and heat coming through windows, can be exhilarating, possibly more so than wall-to-wall synthetic fabrics.

Clear glass allows the bright rays of the sun to shine directly on specific surfaces in the room, leaving others in shadow. Translucent glazing, on the other hand, diffuses light and distributes it more widely, assuring more even heat distribution to many interior surfaces at the same time. This results in more even temperatures and greater heat absorption and storage throughout.

The temperature swings of thermal mass placed in direct sunlight will be about twice the temperature swings of the room itself. Mass shaded from the sun inside the room (such as in north walls) will fluctuate in temperature about half as much as the room. Thus, solar radiated mass stores four times more energy than the shaded mass.

Too little heat storage will allow wide temperature swings and permit overheating which, in turn, wastes heat due to greater heat loss from the house (especially if you open windows to vent that extra heat). Conversely, more mass increases both comfort and the efficiency of the passive system.

The effectiveness of mass also depends on its thickness. The deeper parts of thick walls and floors are insulated by the surface layers and do not store as much heat. Therefore, 100 square feet of 8-inch-thick wall is more effective than 50 square feet of 16-inch-thick wall, even though they both weigh the same.

Provide for thermal mass in the simplest way possible. Otherwise it can be costly and can complicate construction. When used wisely, on-site locally available building materials (gravel, stone, etc.) can be the best kinds of thermal mass. Their use requires less energy than it takes to make and transport brick and concrete.

Rules of thumb for thermal mass

If sunlight strikes directly on the mass (such as a brick floor), each square foot of a window needs roughly 2 cubic feet of concrete, brick, or stone to prevent overheating and to provide heat at night. If sunlight does not strike the mass, but heats the air that in turn heats the mass, four times as much mass is required.

The ability of a material to store heat is rated by its "specific heat," meaning the number of Btus required to raise 1 lb of the material 1°F in temperature. Water, which is the standard by which other materials are rated, has a specific heat of 1.0, which means that 1 Btu is required to raise 1 lb of water 1°F. The 1 lb of water, in turn, releases 1 Btu when it drops 1°F. The specific heat of materials that might be considered for use in the construction of buildings are listed below. The second column of numbers in the table shows the densities of the materials in relation to each other. The material's heat capacity per cubic foot (listed in the third column) was obtained by multiplying its specific heat by its density. Note that the density of water is least among the materials listed but that its heat capacity per cubic foot is still highest because of its high specific heat of concrete (0.2, or 1⁄5 that or water) is partially compensated by its heavy weight and it stores considerable heat (28 Btus per cubic foot for concrete, or about one half that of 62.5 for water). Except for water, the best readily available materials are concrete, brick, and stone.

Specific Heats, Densities, and Heat Capacities of Common Materials

Material	Specific Heat[a]	Density[b]	Heat Capacity[c]
Air (75°F)	0.24	0.075	0.018
Sand	0.191	94.6	18.1
White Pine	0.67	27.0	18.1
Gypsum	0.26	78.0	20.3
Adobe	0.24	106.0	25.0
White Oak	0.57	47.0	26.8
Concrete	0.20	140.0	28.0
Brick	0.20	140.0	28.0
Heavy stone	0.21	180.0	38.0
Water	1.00	62.5	62.5

a) Btus stored per pound per degree change of temperature.
b) pounds per cubic foot.
c) Btus stored per cubic foot per degree change of temperature.

Concrete Floors

Concrete floors are commonly used for storing heat from solar windows. Consider this oversimplified case: A 20 x 40 foot house has a concrete floor 8 inches thick (530 cubic feet). By late afternoon, the slab has been solar heated by 150 square feet of window to an average of 75°F. During the night, the outdoor temperature average 25°F and the indoor air averages 65°F. A well-insulated house may lose heat at a rate of about 200 Btus per hour for each degree of temperature difference (called Delta T, or ΔT) between the oudoors and indoors. The temperature of the slab drops as it loses heat to he house.

The heat lost from the house is the product of the total heat loss rate, the time, and the average temperature difference between indoors and outdoors. In this case, the heat loss during the 15 hour winter night is

(15 hours) x (200 Btus per hour per °F) x (65° - 25°) = 120,000 Btus.

With a heat capacity of about 28 Btus per cubic foot per degree of temperature change, the 530 cubic foot concrete slab stores roughly 15,000 Btus for each degree rise in temperature. For each degree drop in temperature, the slab releases the same 15,000 Btus. If the floor drops 8 degrees, from 75°F to 67°F, it will release just enough heat, 120,000 Btus, to replace the heat lost by the house during the night.

When you calculate mass floor areas, realize that the mass must be left exposed in order to work. Although concrete floors perform well, even the best designed floors for solar exposure very often get covered or shaded by rugs or furniture.

Movable Insulation

Glass loses heat up to 30 times faster than well-insulated walls, so the night-time insulation of glass in winter climates is very important. So is the use of double glazing, which has only half the loss rate of single glass. If the double glazing faces south, it gains more heat than it loses during the winter, virtually anywhere in the country.

In climates of more than 5,000 degree days, the extra cost of low-e glass or triple glazing is usually justified by the energy savings. However, more than three layers of glass is usually not, since each layer also blocks fifteen to twenty percent of the solar energy that passes through the preceding layer.

Multilayered nonglass glazing systems of high transmittance (up to 97 percent), such as Teflon™, can often be used in four or five layers effectively. The reason for this is that they are so clear that an additional layer reduces heat loss significantly, yet blocks very little of the incoming sun.

These thin plastics are not commonly used in home construction; however, for existing homes, thin-film plastics are frequently used in place of glass storm windows.

During the day, when the sun is shining, windows are net energy producers. But since outdoor temperatures are much lower at night, up to three quarters of a window's 24-hour heat loss can be prevented by the proper use of good movable insulating devices, whose heat-flow resistance range from R-4 to R-10.

Moveable Insulating Devices

- Sheets of rigid insulation manually inserted at night and removed in the morning.

- Framed and hinged insulation panels.

- Roller-like shade devices of one or more sheets of aluminized mylar, sometimes in combination with cloth and other materials.

- Sun-powered louvres which automatically open when the sun shines and close when it doesn't.

- Mechanically-powered systems which use blowers to fill the air space between two layers of glazing with insulating beads at night.

A window loses heat to the out-of-doors in proportion to the temperature of the air space between the window and the insulation provided. A loose-fitting insulating shutter will allow room air into that space and diminish the insulating effect. Therefore, a snug fit and sealed edges are important.

A few cautionary words: Sun shining on an ordinary window covered on the inside by a tight insulating shade can create enough thermal stress to break the glass. A white or highly reflective surface facing the glass is the best solution, but not foolproof. Also, moisture can condense at night on the cold window glass facing the insulation, causing deterioration of the wooden frames. Tight-fitting insulation is the best solution for preventing excessive condensation. Otherwise, provisions for collecting and draining the condensation may be necessary. Also, remember to conform to all codes; don't use insulation materials that are flammable or in other ways hazardous without protecting them properly.

Energy Savings from Movable Insulation

To determine the annual energy savings using movable insulation, first find the difference between the U-value of the window as it is and the U-value of the window using movable insulation. Then multiply this difference times the number of degree days where you live times 24 hours per day.

For example, suppose that an insulating panel with a heat flow resistance of R-10 is being considered for windows in Minneapolis with two layers of glass. Assume that the insulation will be in place an average of 12 hours per day. The U-value of the glass is 0.55 Btu/sq ft/hr/°F (from Appendix 4). The U-value for the insulated window system is 0.24. The difference between the two is 0.31 (0.55 − 0.24). Minneapolis averages 8,382 degree days per year (from Appendix 5). Therefore

annual energy savings = (0.31 Btu/sq ft/hr/°F)

x (8,382 deg days/yr) x (24 hr/day)

= 63,362 Btu/sq ft/yr

For a 10-square-foot window, the savings is roughly equivalent to the heating energy obtained from 180 kwh of electric resistance heating, from 10 gallons of oil burned in most furnaces, or from 10 square feet of an active solar collector of average design. A tight-fitting shutter also reduces heat loss due to air leakage around the window frame, making the above savings a conservative estimate.

In and Out the Windows

Today's windows are smarter than ever before. The old choice of single versus double versus triple glazing has been replaced by double glazing versus a relatively new glazing called "low-e" for low-emissivity. These double (and triple) glazings have a thin metal coating that allows short-wave solar energy to pass through, but blocks most of the long- wave thermal energy that tries to escape back out. These glazings have almost the same insulating value as triple glazing, but weigh much less, resulting in lower shipping costs and easier installation.

Although triple glazing is still available, it has all but been replaced by low-e glazings in most applications. Instead of filling the airspace between the panes of glass with air, many manufacturers are using a higher density gas such as Argon, which has a lower conductivity than air. Other gases, such as krypton and sulfur hexafluoride, are being experimented with. Soon to hit the market are glazings made with liquid crystal properties that change from clear to opaque with the addition of an electric charge.

The low-emissivity coating is virtually invisible, but helps keep the inner layer of glass at a higher temperature. This results in less radiant heat loss from your body to the glass, making you feel more comfortable.

The higher glass temperature also helps control condensation on the glass. New edge seals on windows also help control condensation. Look for windows that have a "thermal break," where a less-conductive spacer is used to separate the inside and outside glazing layers. This will reduce edge-of-glass condensation. (In new construction, recessing the window into the wall—especially in thicker wall construction—will reduce condensation by allowing indoor warm air to naturally thermosiphon up over the window glass, raising the temperature of the inner glass surface.)

An added bonus of the low-e glass is that it blocks more ultraviolet light than clear glass, slowing down the fading of furniture, rugs and curtains.

The new low-emissivity glazings can reduce heat loss over their clear counterparts by 7 to 12 percent, while only reducing solar heat gain 5 to 18 percent. Installing argon- filled low-e windows can reduce heat loss by as much as 25 percent over double glazing, but cut solar gain no more than their low-e counterpart.

In south-facing windows intended for direct solar gain, look for windows that have a high shading coefficient and a high R-value (see Table 1). For north-, east-, and west-facing windows, choose windows with low-e with Argon, and an R-value of 3 to 3.5. Add window insulation where possible, and keep it in place as long as you can (see Table 2).

Table 1 Glazing Properties

	R-value*	SC	VT	LE
Pella Low-E Insulated	2.40	0.83	0.74	0.89
Single Glazing	1.11	1.00	0.90	0.90
PPG Sungate	3.23	0.86	0.78	0.91
Clear Double Glazing (1/4)	1.85	0.89	0.81	0.91
Clear Double Glazing (1/2)	2.04	0.89	0.81	0.91
LOF Low-E Hardcoat	2.86	0.82	0.75	0.91
LOF Hardcoat w/Argon	3.33	0.82	0.75	0.91
Clear Triple Glazing	2.38	0.80	0.74	0.93
Marvin Double Glazing	2.13	0.84	0.82	0.98
Heat Mirror 88	2.90	0.70	0.71	1.01
Anderson HP w/Argon	3.20	0.77	0.79	1.03
Marvin Double Low-E	2.78	0.71	0.74	1.04
Marvin Double Low-E w/Argon	3.13	0.71	0.74	1.04
Cardinal Low-E Softcoat	2.56	0.71	0.77	1.08
Cardinal Low-E Squared w/Argon	2.78	0.71	0.77	1.08
Pella Low-E Insulated w/Argon	2.90	0.73	0.79	1.08
Cardinal Sun-Blocking Low-E	2.70	0.40	0.43	1.08
Cardinal S.B. Low-E w/Argon	3.13	0.40	0.43	1.08
Heat Mirror 66	3.00	0.50	0.56	1.10
Heat Mirror 55		0.41	0.47	1.15
Hurd Insol-8	4.60	0.52	0.62	1.19
Cardinal Low-E Softcoat	2.86	0.49	0.69	1.41
Cardinal Low-E2 w/Argon	3.33	0.49	0.69	1.41
Weather Shield Tripane	2.78	0.85		

* R-values are for the total window unit, evaluated using Window 3.1 by Lawrence Berkeley Laboratory and 1989 ASHRAE *Handbook of Fundamentals.*

Table 2 Average Winter U-Values* for Windows with Insulating Covers

	No insulation	R-4 cover	R-4 cover[a]	R-4 cover[b]	R-10 cover	R-10 cover[a]	R-10 cover[b]
Single glazing	1.11	0.20	0.43	0.50	0.092	0.35	0.43
Double glazing 1/2" Airspace	0.49	0.17	0.25	0.27	0.083	0.18	0.22
Triple glazing 1/4" Airspace	0.42	0.16	0.22	0.24	0.081	0.17	0.20
Low-E Hardcoat 1/2" Airspace	0.40	0.15	0.21	0.24	0.080	0.16	0.19
Low-E Hardcoat 1/2" Argon	0.39	0.15	0.21	0.23	0.079	0.16	0.18
Low-E Softcoat 1/2" Airspace	0.38	0.15	0.21	0.23	0.079	0.15	0.18
Low-E Softcoat 1/2" Argon	0.36	0.15	0.20	0.22	0.078	0.15	0.17
Heat Mirror™ 1-susp. films	0.34	0.14	0.19	0.21	0.077	0.14	0.16
Heat Mirror™ 2-susp. films	0.22	0.12	0.14	0.15	0.069	0.11	0.12

*U-value = Btu/hr/sq ft/°F. Values are for the total window unit, evaluated using Window 3.1 by Lawrence Berkeley Laboratory and 1989 ASHRAE Handbook of Fundamentals.

[a]Cover in place 16 hrs/day (3/4 of deg days).

[b]Cover in place 12 hrs/day (2/3 of deg days).

Buying by Type

Casements, awnings, hoppers—those windows that press shut against the frame—have half the window infiltration of those windows that glide (sliders, gliders). Fixed windows have the lowest infiltration. Large expanses of south-facing glass will lose less energy if they are fixed. Situate operable windows for maximum ventilation in each room.

Insulating Values

In 1990, a new calculation was adopted for U-values that takes into the window frame, showing insulating values that are around 25% lower than pre-1990 numbers, when R-values were only measured for the "center-of-glass." So make sure you're working with recent window catalogues and are comparing apples-to-apples. When comparing window values, make sure they have been evaluated using Window 3.1 by Lawrence Berkeley Laboratory and the 1989 ASHRAE *Handbook of Fundamentals*.

Costs

Buying a window with a low-emissivity coating can add around 4 to 12 percent to the cost of a window. But the added expense is so cost effective that some major manufacturers are no longer even offering clear glass in some areas. The added cost of Argon ranges from nothing to pennies per square foot.

Passive Gain

The shading coefficient (SC) of a window tells you how much solar heat can be gained, compared with one layer of clear single glazing (1.00). The higher the SC, the better solar collector the glazing will make.

In Northern regions, increase gain by picking a SC higher than 80 percent. In Southern states, reduce gain but not visible light, and pick a window with a SC lower than 60 percent—but don't go too low. You should also pay attention to the visible light transmittance (see Table 1). Low visible light transmittance, such as in reflective glass, can drastically reduce natural lighting. So look for a balance between the two. If you divide the visible transmittance by the shading coefficient, you get a number called the "Luminous Efficacy" constant or LE. The higher this number, the more suitable the window is for southern climates. The lower the number, the more suitable for maximum solar gain in northern climates.

Solar Chimneys

A solar chimney is an air-heating solar collector that runs automatically, on sun power alone. Of all the passive heating systems, it loses the least heat when the sun is not shining. Except for solar windows, solar chimneys (also called convective loops or TAPs, thermosiphoning air panels) are the most common solar heating systems in the world.

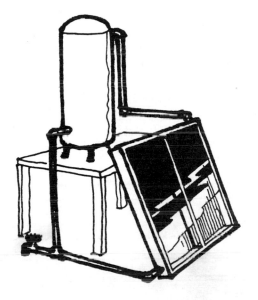

Variations on the design are used to heat water for domestic purposes. Hundreds of thousands of pumpless "thermosiphoning" (heat convecting) solar water heaters have been used for decades. In fact, convective loop water heaters were patented in 1909. By 1918, 4,000 such heaters were in operation in Southern California.

Solar chimney wall-mounted collectors can complement south-facing windows in supplying additional solar heat directly to both new and existing houses. In conventional wood-framed houses, up to 25 percent of the heat can be supplied by combining solar windows and solar chimneys without supplemental thermal storage.

Basic System Design

A solar chimney wall collector is similar to a flat-plate collector used for active systems. A layer or two of glass or plastic covers a black absorber. Air may flow in a channel either in front of or behind the absorber, depending on the design. The air may also flow through the absorber if it is perforated. The collector is backed by insulation.

In the exploded view shown here, the collector is mounted on, or made a part of, the insulated wall. Openings at the bottom and top of the wall permit cooler air from the house to enter the hot collector at the base of the wall, to rise as the sun heats it, and to vent back into the building near the ceiling.

The slow-moving collector air must be able to come in contact with as much of the absorber's surface area as possible without being slowed down too much. In fact, the amount of heat transferred from the absorber to the flowing air is in direct proportion to he heat-transfer capabilities of the absorber and the speed of the air flow by or through it.

Up to six layers of expanded metal lath are used in some absorbers. In these, the air rises in front of the lath, passes through it, and leaves the collector through a channel behind the lath. Flat or corrugated metal is also used, but it does not transfer heat as well. However, the air-flow channel in this case need not be as deep. The metal should be placed in the center of the channel, if possible, so that air flows on both sides. This is more difficult to do and requires two glazing layers instead of one.

Construct the collectors carefully and insulate them well, particularly the upper areas that are likely to be hottest. Avoid insulations or glazings that will melt. If the collector's flow should be blocked for some reason on a sunny day, its temperature can reach over 300°F. Wood construction is usually satisfactory, but be sure to provide for wood shrinkage and for the expansion and contraction of materials as their temperatures fluctuate.

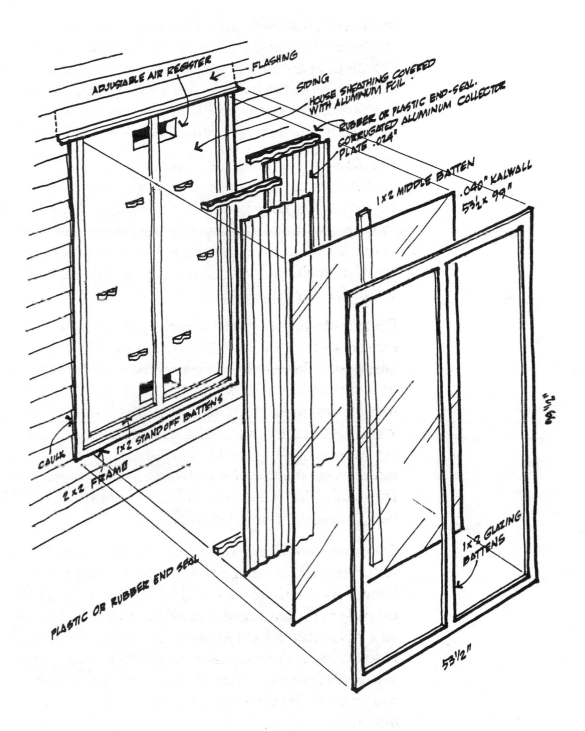

ADJUSTABLE AIR REGISTER

FLASHING

SIDING

HOUSE SHEATHING COVERED WITH ALUMINUM FOIL.

RUBBER OR PLASTIC END-SEAL.

CORRUGATED ALUMINUM COLLECTOR PLATE .024"

1 x 2 MIDDLE BATTEN

.040" KALWALL 53½ x 99"

84½"

CAULK

1 x 2 STANDOFF BATTENS

2 x 2 FRAME

1 x 2 GLAZING BATTENS

PLASTIC OR RUBBER END SEAL

53½"

Systems with Heat Storage

Collectors that are large enough to supply more than 25 percent of the desired heat require heat storage. The storage in an air system is usually a large bin or rocks. It must be designed to maximize heat transfer from the air stream to the rocks without noticeably slowing the air flow. Rocks with small diameters (three to six inches) have large amounts of surface area for absorbing heat, and yet allow passages for air flow. The rocks should be roughly the same size (that is, don't mix one inch with four inches) or most of the airways will be clogged. Storage should contain at least 200 pounds of rock (one and one half cubic feet) per square foot of collector. As shown in the diagram, storage should be located above the collector but below the house. This permits solar-heated air to rise into the house, and cooler air to settle in the collector.

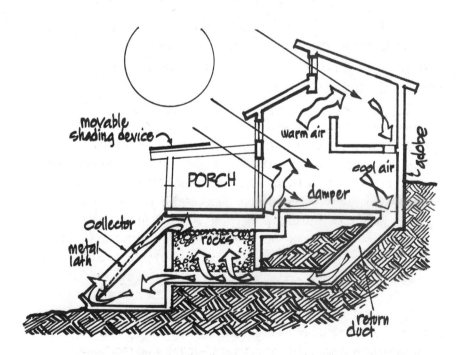

Air flow

As with active collectors, the slower the air flow, the hotter the absorber, and the greater the heat loss through the glazing. This results in a lower collector efficiency. Good air flow keeps the absorber cool and transports the maximum possible amount of heat into the house. Flow channels should be as large as possible, and bends and turns in the ducts should be minimized to prevent restriction of air flow.

Conventional air heating collectors use fans and have air channels only one-half to one inch deep. Without fans, air channels in convective loop collectors range from two to six inches deep.

Convective flow of air is created by a difference in temperature between the two sides of the convective loops, for example, between the average temperature in the collector and the average temperature of the adjacent room. It is also affected by the height of the loop. The best air flow occurs when the collector is hottest, the room is coolest, and the height of the collector is as tall as possible.

The cross-sectional area of the rock bed receiving air from the collector should range from one half to three quarters the surface area of the collector. The warm air from the collector should flow down through the rocks, and the supply air from storage to the house should flow in the reverse direction. Optimum rock size depends on rock bed depth. One authority recommends gravel as small as one inch for rockbeds two feet deep, and up to six inches for depths of four feet. For best heat transfer in active systems, bed depths are normally at least 20 rock diameters. That is, if the rock is four inches in diameter, the bed should be at least six and one half feet deep in order to remove most of he heat from the air before it returns to the collector. This should be considered a maximum depth for convective-loop rockbeds.

The vertical distance between the bottom air inlet and the top air outlet should be at least six feet to obtain the necessary effect. It should be tilted at a pitch of not less than a 45° angle to the ground, to allow for a good angle of reception to the sun and for the air to flow upward.

Reverse air convection

In an improperly designed system, reverse air convection can occur when the collector is cool. A cool collector can draw heat from the house or from storage. Up to 20 percent of the heat gained during a sunny day can be lost through this process by the following day.

The are three primary methods of automatically preventing reverse convection. One is to build the collector in a location below the heat storage and below the house, so that cold air will sink to the bottom and thermosiphoning will cease. A second is to install backdraft dampers that automatically close when air flows in the wrong direction. One such damper is made of lightweight, thin plastic film. A lightweight "frisket" paper used in the photography industry has also been used successfully. Warm air flow gently pushes it open. Reverse cool air flow causes the plastic to fall back against the screened opening, stopping air flow. Ideally, both top and bottom vents should be equipped with such dampers.

The third method of reducing reverse convection is to place the intake vent slightly lower than the outlet vent near the top of the wall. The back of the absorber is insulated and centered between the glazing and the wall. Inlet cool air from the ceiling drops into the channel behind the absorber. The solar heated air rises in the front channel, drawing cooler air in behind it. The warmed air enters the room at the top of the wall. When the sun is not shining, the air in both channels cools and settles to the bottom of the "U-tube." Only minor reverse convection occurs. Because of the longer air-travel distances involved, the U-tube collector will not be as efficient, aerodynamically, as the straight convective loop. It will also be more expensive to build.

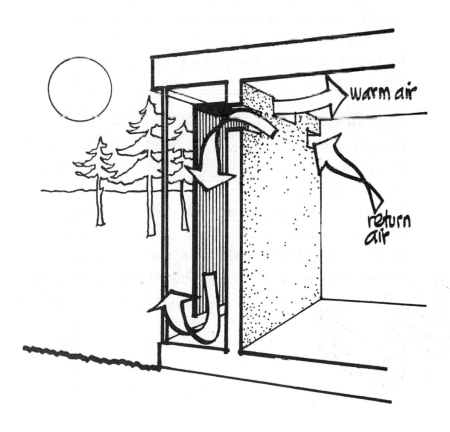

Performance

Performance of a solar chimney depends largely on delicate, natural convection currents in the system. Therefore, proper design, materials, and construction are important. In a well-built collector, air flow can be low to nonexistent at times of little or no sun, but will increase rapidly during sunny periods. Average collection efficiency is similar to that of low temperature, flat-plate collectors used in standard active system designs.

Collector Area

Only a small collector area is needed to heat a house in the spring and fall when the heating demand is low. Additional collector area provides heat over a fewer number of months, only during the middle of the heating season. Therefore, each additional square foot of collector supplies slightly less energy to the house than the previous square foot.

The useful amount of heat supplied from a solar collector ranges from 30,000 to 120,000 Btus per square foot per winter. The high numbers in this range are for undersized systems in cold, sunny climates. The low numbers are for oversized systems or for very cloudy climates.

In cold climates of average sunshine, such as Boston, Massachusetts, 50,000 to 80,000 Btus per square foot per heating season are typical, depending on the size of the solar system. (For comparison, roughly 80,000 Btus are obtainable from a gallon of oil.)

Solar Walls

Solar windows let sunlight directly into a house. The heat is usually stored in a heavy floor or in interior walls. Thermal storage walls, as solar walls are often called, are exactly what their name implies—walls built primarily to store heat. The most effective place to build them is directly inside solar windows, so that the sunlight strikes the wall instead of directly heating the house. A directly sun-heated wall gets much hotter, and thereby stores more energy, than thermal mass placed elsewhere.

These "solar walls" conduct heat from their solar hot side to their interior cooler side, where the heat then radiates to the house. But this process takes a while. In a well-insulated house, a normal number of windows in the south wall will admit enough sun to heat the house during the day. Thermal storage walls will then pick up where the windows leave off and provide heat until morning.

South-facing windows with an area of less than 10 percent of the floor area of the house are probably not large enough to provide enough heat during the day. If this is the case, vents could be added at both the base and the top of a solar wall. The wall can then provide heat to the house during the day, just as solar chimneys do. Although the vents need be only 10 to 12 square inches for each lineal foot of wall, they can add cost and complication. Therefore, it is best not to use them unless heat is needed during daylight hours.

73

Thermal storage walls with vents are called Trombe Walls, after Dr. Felix Trombe, who in the early 1960s in France built several homes with this design. (Actually, the concept was originated and patented by E. L. Morse of Salem, Massachusetts, in the 1880s. His walls, complete with top and bottom dampers, used slate covered by glass.)

One type of thermal storage wall uses poured concrete, brick, adobe, stone or solid (or filled) concrete blocks. Walls are usually one foot thick, but slightly thinner walls will do, and walls up to 18 inches thick will supply the most heat. Further thicknesses save no additional energy.

Containers of water are often used instead of concrete. They tend to be slightly more efficient than solid walls because they absorb the heat faster, due to convective currents of water inside the container as it is heated. This causes immediate mixing, and quicker transfer of heat into the house than solid walls can provide. One-half cubic foot of water (about four gallons) per square foot of wall area is adequate, but unlike solid walls, the more water in the wall, the more energy it saves.

The main drawback of solar walls is their heat loss to the outside. Double glazing (glass or any of the plastics) is adequate for cutting this down in most climates where winter is not too severe (less than 5,000 degree days: Boston, New York, Kansas City, San Francisco). Triple glazing or movable insulation is required in colder climates.

Installation costs are affected by local construction practices, building codes, labor rates and freight rates. Walls made of poured concrete and masonry block are less expensive in areas of the country where these materials are commonly used. The exterior glazing can be reasonable in cost if an experienced subcontractor is available, or if materials can be obtained inexpensively through local suppliers. Operating costs for solar walls are zero, and little or no maintenance is required.

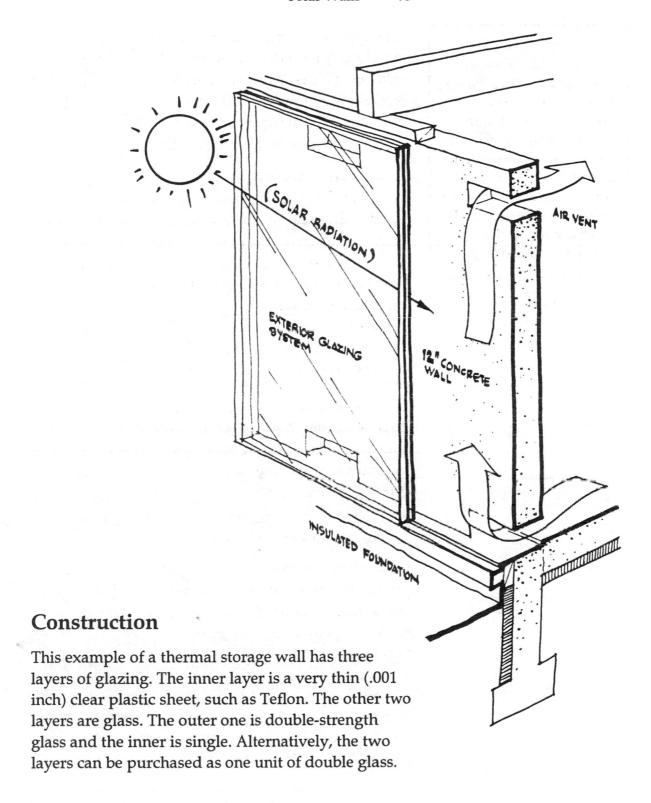

(SOLAR RADIATION)

EXTERIOR GLAZING SYSTEM

12" CONCRETE WALL

AIR VENT

INSULATED FOUNDATION

Construction

This example of a thermal storage wall has three layers of glazing. The inner layer is a very thin (.001 inch) clear plastic sheet, such as Teflon. The other two layers are glass. The outer one is double-strength glass and the inner is single. Alternatively, the two layers can be purchased as one unit of double glass.

Mount the entire glazing system one or two inches away from the wall. If the wall has vents, mount the glazing 3 to 4 inches away to allow for adequate air flow. Be sure to provide for the removal of cobwebs from the air space and for cleaning and replacement of all glazing components. If you use aluminum rather than wood for framing and mounting the glass, place wood or other insulating material between the aluminum and the warm wall as a thermal separator. The glazing should extend above and below the face of the storage wall and be fully exposed to the sun. Glazing must be airtight and water resistant; it is the weather skin of the building.

The wall itself is of concrete, 12 inches thick, and of any height or width. It is either poured-in-place concrete, solid concrete-block masonry, or concrete blocks filled with cement mortar. Use regular stone aggregate in the concrete (about 140 pounds per cubic foot). Lightweight aggregates should not be used, since they do not store as much heat.

Unless the wall has to do structural work as well (such as supporting another wall or the roof), the concrete mix can be a relatively inexpensive one. When, as in many cases, the wall also supports the roof, reinforcing steel and structural anchors can be added without altering the wall's solar performance.

In general, treat the juncture between the inner storage wall and the foundation floors, adjacent side walls, and roof, as normal construction. However, make sure that house heat cannot easily escape through masonry or metal that is exposed to the weather. For example, foundations directly below glazed thermal storage walls should be doubly well-insulated from the ground.

If you install vents in the wall, use backdraft dampers to prevent reverse air circulation at night. (See the chapter on solar chimneys for an example of a damper you can build.) Place the vents as close to the floor and ceiling as possible. Openings may be finished with decorative grills or registers. Such grills will keep inquisitive cats and tossed apple cores out of the airway, too!

Any interior finish must not prevent heat from radiating to the room. Just seal and paint the wall any color, or sandblast or brush the surface to expose the stone aggregate. A plaster skim coat can be used. Sheet materials, such as wood or hardwood paneling, should not be used. Use gypsum board only if excellent continuous contact between the board and the wall can be obtained—a difficult if not impossible task. Remember that many architects and interior designers regard natural concrete as an acceptable and attractive interior surface material.

Cleanse the solar (outer) surface of the wall with a masonry cleaner prior to painting with virtually any dull-finish paint. Although dark brown and green have been used, flat black paint is preferred for maximum heat absorption.

An alternative to the solid concrete wall is vertical solar louvers, a set of rectangular columns oriented in the southeast-northwest direction (see plan at window). They admit morning light into the building and store much of the heat from the afternoon sun. The inside of the glass is accessible for cleaning, and movable insulation can be easily installed between the glazing and the columns.

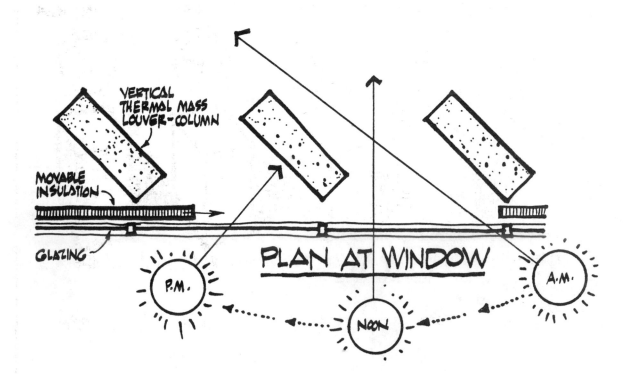

VERTICAL THERMAL MASS LOUVER-COLUMN

MOVABLE INSULATION

GLAZING

PLAN AT WINDOW

P.M.

NOON

A.M.

The Heating Effect

Heat loss through solar walls, even after days of cloudy weather, is not much worse than through conventional walls. The overall U-value of solar walls is 0.23. Due to their solar gains, they are net energy producers. The temperatures of the heat supplied by solar walls is generally more moderate than that supplied by conventional heating systems. The lower temperatures tend to be more comfortable and less drying. Vented solar walls provide air at the ceiling level at 90°F to 100°F at air flows of approximately one cubic foot per minute per square foot of wall area. The interior surface of a 12-inch thick wall reaches its warmest temperature in the early evening and then coasts from there on by releasing stored heat. Temperatures range from 65°F to 85°F over the course of a day. The interior surface of a 24-inch wall peaks in temperature about 8 hours later. It doesn't get nearly as warm as thinner walls, but it provides heat more evenly for a longer period of time.

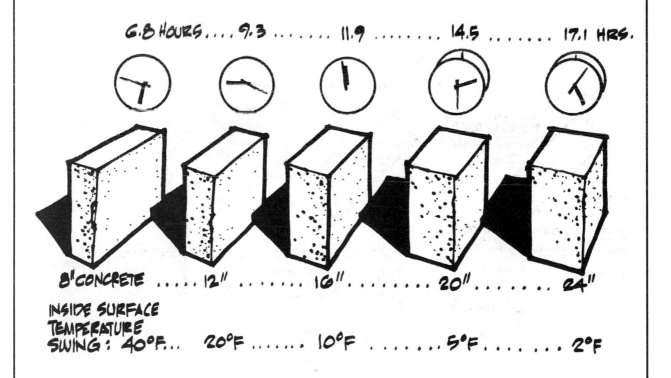

6.8 HOURS.... 9.3 .... 11.9 .... 14.5 .... 17.1 HRS.

8" CONCRETE ..... 12" .... 16" .... 20" .... 24"

INSIDE SURFACE
TEMPERATURE
SWING: 40°F... 20°F .... 10°F .... 5°F .... 2°F

Summer Shading

A solar wall will supply a small amount of heat through the summer and have an effect on cooling bills. Shading the wall with an overhang, an awning, or a tree is the most effective method of reducing its exposure to direct sunlight. A cloth or canvas draped over the wall is also effective. Some people choose to place vents in the framing at the top of the wall. The vents are open during the summer, permitting the heat from the wall to escape to the outside. Their primary shortcoming is that they are prone to leakage during the winter. This air leak can have a significant effect on the performance of the wall during the winter.

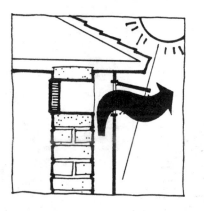

Converting Your Existing Home

Solar walls are more difficult to add to an existing home than are solar windows, solar chimneys and solar roofs. Uninsulated brick, stone, adobe or block walls are candidates for conversion if they are unshaded during the winter and if they are oriented in a southerly direction (within 30 degrees east or west of due south). Solid walls are more effective than walls that have air spaces, as is often found in brick walls comprised of two layers.

If possible, conventional interior finish materials should be removed from the inner surface of the wall. Openings for vents are usually very difficult to make in walls that are candidates for conversion to solar walls. If, however, your windows are not large enough to supply all the heat you need during the day, and if heat is more important during the day than during the night (as for example, in an office building or a school), install vents of the sizes and in the proportions described earlier in this chapter for new walls.

Paint the wall, and then cover it with the glazing system appropriate to your climate. For unvented walls, cover the wall first with an inexpensive sheet of plastic to bake the solvents out of the paint. When the plastic is coated with a thin film, remove the plastic and proceed with the installation of the permanent glazing system.

Solar Roofs

Solar roofs, often called thermal storage roofs, are similar to storage walls. Waterbed-like bags of water exposed to sunlight collect, store and distribute heat. This heat passes freely down through the supporting ceiling to the house, gently warming it. In the summer, heat rises through the ceiling into the water, cooling the house. Then at night the water is cooled by the radiation of its heat to the sky. Movable insulation covers the ponds at night in winter to trap heat inside and during the day in summer to shade the ponds while the sun is shining.

Generally, solar roof ponds are 8 to 12 inches deep. Roof ponds are always flat, but in northern buildings the glazing is often sloped to he south to capture the sun's low rays as well as to shed snow. Under the sloped glass, the walls are well-insulated and faced with materials that reflect the sunlight into the ponds.

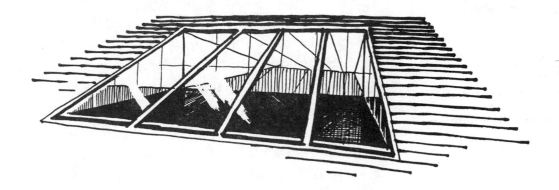

Solar roof ponds maintain very stable indoor temperatures. During the winter and summer, temperatures typically fluctuate between 66°F and 73°F, while the outdoor average daily temperature fluctuates between 47°F and 82°F throughout the entire year.

Not many solar roofs have been built, and there is limited information on the design, cost, performance and construction details of thermal storage roofs. However, they offer tremendous potential for reducing heating and cooling bills.

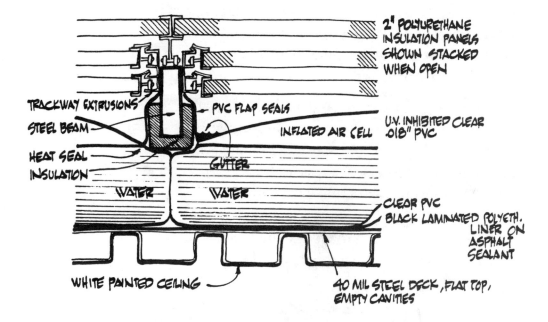

2" POLYURETHANE INSULATION PANELS SHOWN STACKED WHEN OPEN

TRACKWAY EXTRUSIONS

STEEL BEAM

PVC FLAP SEALS

INFLATED AIR CELL

U.V. INHIBITED CLEAR .018" PVC

HEAT SEAL INSULATION

GUTTER

WATER   WATER

CLEAR PVC
BLACK LAMINATED POLYETH. LINER ON ASPHALT SEALANT

WHITE PAINTED CEILING

40 MIL STEEL DECK, FLAT TOP, EMPTY CAVITIES

Roof Detail

This cross section is of the solar roof system used in a house designed by Harold Hay in the mild climate of Atascadero, California. The entire 1100-square-foot ceiling is covered with eight inches of water sealed in clear, UV-inhibited, 20-mil polyvinylchloride water bags.

Underneath this 53,600 pounds of water is a layer of black polyethylene to help absorb solar radiation near the bottom of the bags. Additionally, an inflated clear plastic sheet above the water bags enhances the "greenhouse effect" during the heating season. This air cell is deflated in the summer months to permit radiational cooling. A 40-mil steel deck roof supports the waterbags and provides good heat transfer to and from the living space.

Above the roof ponds, a system of movable insulating panels is mounted on horizontal steel tracks. The insulation is two inches of rigid polyurethane faced with aluminum foil. The panels are moved by a 1/6 horsepower motor, operating about 10 minutes per day.

The Atascadero house is 100 percent solar heated and cooled, and it has no other source of heating or cooling. Occupants have found that the heating and cooling system provides "superior"" comfort compared with conventional systems.

How the Sun can Cool

"What do you mean by *solar* cooling?" That's a good question. In fact, of the several passive solar cooling techniques discussed here, only one of them can be labeled strictly "solar," and even at that only if you stretch the definition. But all the techniques use passive or natural cooling that requires no pumps or fans for its operation. Many of them are based on plain old common sense.

Fortunately, in virtually all climates houses can be designed to provide ideal comfort without mechanical air conditioning. This is not to say it is easy to do everywhere. In really hot and humid climates where electricity is still relatively inexpensive, it just might not be worth the extra effort. It's at least nice to know such things are possible.

By far, the first and most important step in cooling is to keep sunlight from falling on your house, so first we'll discuss "solar control." We'll also discuss natural cooling by ventilation, evaporation, sky radiation and nesting of buildings into the earth.

Shading

Usually the easiest, most inexpensive and effective way to "solar" cool your house is to shade it—keep the sun from hitting your windows, walls and roof. In fact, where summer temperatures average less than 80°F, shading may be all you need to stay cool.

Most of the things you do to reduce winter heat loss also reduce summer heat gain. For example, heavily insulated walls keep out summer heat, so shading them is not as important as if they were poorly insulated. Facing windows south to catch the winter sun, and minimizing east and west windows too reduce heat loss, are also important steps in solar control. South windows admit less sun in the summer than they do in the winter, while east and west windows can turn houses into ovens. A light-colored roof may be 60 to 80 degrees cooler than a dark roof because it reflects light.

In really hot climates, uninsulated walls and roofs should be shaded. But, although this is important, shading windows is far more important. Overhangs and awnings work well. Unfortunately, fixed overhangs provide shading that coincides with the seasons of the sun rather than with the climate. The middle of the sun's summer is June 21—the longest day, when the sun is highest in the sky. But the hottest weather occurs in August, when the sun is lower in the sky.

A fixed overhang designed for optimal shading on August 10, 50 days after June 21, also causes the same shading on May 3, 50 days before June 21. The overhang designed for optimal shading on September 21, when the weather is still somewhat warm and solar heat gain is unwelcome, also causes the same shading on March 21, when the weather is cooler and solar heat is welcomed.

Shading from deciduous vegetation more closely follows the climatic seasons, and therefore the energy needs of houses. On March 21, for example, there are no leaves on most plants in north temperate zones

---

**Reducing the Need
for New Power Plants**

Natural cooling can significantly reduce peak cooling power loads. With natural cooling, the size and use of backup conventional electrical air conditioners is reduced. This means the demand for power is reduced, and so is the need for new power plants. For example, approximately 50 square feet of west window, unshaded from the sun, needs a one-ton air conditioner and approximately two kilowatts of electrical generating capacity to offset the solar heat that comes through it. Shading the windows, or facing them in other directions, can reduce those peak power demands created during the summer months. Thermal mass can absorb heat during the day, delaying the need for cooling until after peak demand hours. At that later time the need will not be as great and the air conditioner can be smaller and cheaper to run.

and the bare branches readily let sunlight through. On September 21, however, those same plants are still in full leaf, providing needed shading.

Operable shades are even more versatile and adaptable to human comfort. The most effective shades are those mounted on the outside of a building. However, most exterior operable shades do not last very long. Nesting animals, climbing children, wind and weather will see to that. Inside shades last longer, but few are as effective as outside shades. Once the sunlight hits the window glass, half the cooling battle is lost.

East and west glass is difficult to shade because the sun in the east and west is low in the sky in both summer and winter. Overhangs do not prevent the penetration of sunlight through east and west windows during the summer much more than they do during the winter. Vertical louvers or other vertical extentions of the building are the best means of shading such glass.

A significant source of technical detail on shading can be found in the *ASHRAE Handbook of Fundamentals* by the American Society of Heating, Refrigerating, and Air Conditioning Engineers, New York.

Ventilation

The movement of room temperature air, or even slightly warmer air across our skin, causes a cooling sensation. This is because of the removal of body heat by convection currents and because of the evaporation of perspiration.

The most common way to cool a house with moving air, without using mechanical power, is to open windows and doors. Do not forget this simple concept—natural ventilation—and do not underestimate its cooling effect. Low, open windows let air flow through the lower part of the room where people are, rather than near the ceiling.

Houses that are narrow and face the wind, or that are T- or H-shaped, trap breezes and enhance cross ventilation through the house. When all else fails, open or screened porches located at or near the corners of houses, can capture soft, elusive breezes as they glide around the house.

The "stack" or "chimney" effect can be used to induce ventilation even where there is no breeze. Warm air rises to the top of a tall space, where openings naturally exhaust the warm air. Openings at floor level let outdoor air in. Natural ventilation can be further induced by the use of cupolas, attic vents, belvederes, wind vanes and wind scoops.

If your summer nights are a lot cooler than days, build your house of heavy materials. Cool the house at night by natural ventilation and the thermal mass will keep it cool during the day.

Natural ventilation can be affected by land planning.
Natural breezes should not be blocked by trees,
bushes or other buildings. Shade trees should be
selected so that branches and leaves are as high
above the house as possible to allow a breeze to
enter below them. The shape of your house, proper
clustering of buildings, and other landscaping
features, such as bushes and fences, can funnel and
multiply natural breezes.

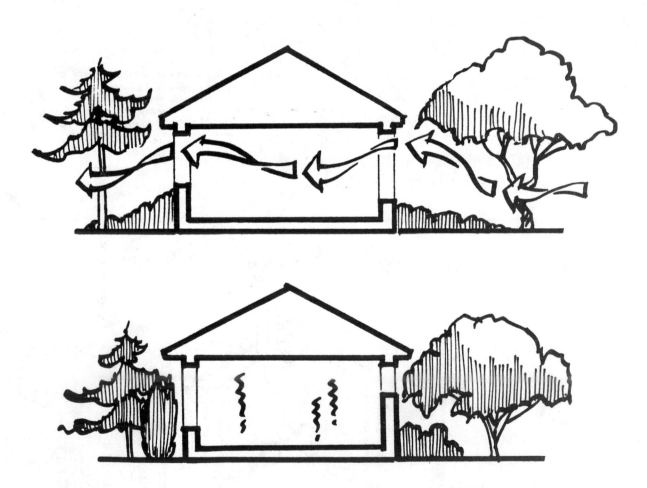

Here, a solar collector exhausts its hot air to the outdoors by natural convection and pulls house air through itself, providing ventilation. The solar collector is very similar to the solar chimney convective loop collectors discussed earlier. Many variations of this "solar chimney" have been used widely in the past and are being developed again today. In some active solar systems using heated air (rather than a liquid), the collectors vent hot air to the outside during sunny summer weather, pulling house air through themselves, with or without the use of blowers.

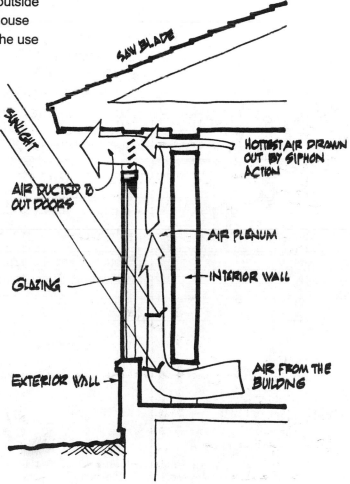

SAW BLADE

SUNLIGHT

AIR DUCTED TO OUT DOORS

HOTTEST AIR DRAWN OUT BY SIPHON ACTION

AIR PLENUM

GLAZING

INTERIOR WALL

EXTERIOR WALL

AIR FROM THE BUILDING

Determining Ventilation Air Flow

In stack-effect ventilation, air flow is maximized by the height of the stack and the temperature of air in the stack. Air flow is proportional to the inlet area and to the square root of the height times that average temperature difference as follows:

$$Q = 540A \sqrt{h(T_1 - T_2)},$$

where
Q = the rate of air flow, in cubic feet per hour;
A = the area of the inlets, in square feet;
h = the height between inlets and outlets, in feet;
T1 = the average temperature of the air in the "chimney," and
T2 = the average temperature of the return air (normally just the outside temperature).

It is better to add heat (presumably using a passive air heating collector) at the bottom of the chimney or stack than at the top. In this way the entire column of air in the chimney is hot, creating the desired bouyancy to cause the air to flow.

If outlet sizes are appreciably different from inlet sizes, the above expression must be adjusted according to the following ratios:

Area of Outlets Area of Inlets	Value to be substituted for 540 in above expression
5	745
4	740
3	720
2	680
1	540
3/4	455
1/2	340
1/4	185

This information is from *Design with Climate* by Victor Olgyay, Princeton University Press.

Evaporation

Our skin feels cooler if it's wet when the wind hits it than when it's dry due to the evaporation of moisture. The same process can cool air effectively, especially in dry climates. There are many ways to evaporate water. The most effective are by having large water surfaces and by agitating, spraying or moving water to get large surface contact with the air. For example, large shallow ponds provide this large surface contact. Moving streams and sprays from water fountains increase turbulence and, thus, surface contact.

Evaporative cooling can also keep roofs cool. Roof sprays and ponds have been used successfully in hot climates such as in Arizona and Florida. Wind-flows across roof ponds should be enhanced, if possible, to encourage evaporation. This can be done mechanically, but careful design of roof shapes can help speed the natural flow of air across the ponds.

The transpiration of indoor plants has a cooling effect. So do interior pools and fountains. Fans can add enormously to the evaporative effect and have been used successfully in "swamp" coolers and other evaporative coolers.

If you live in a humid climate, however, do not expect evaporative cooling to help you very much. In fact, it is likely to make your humidity problems worse.

Radiational Cooling

Thermal energy is constantly being exchanged between objects that can "see" each other. More energy radiates from the warmer object to the cooler object. The sun radiates heat to the earth, and the

earth radiates considerable heat to the upper atmosphere at night.

The most effective radiant cooling surface is horizontal, facing straight up. Obstructions such as trees and walls reduce such cooling. A vertical surface with no obstructions yields less than half of the radiant cooling of an unobstructed horizontal surface.

Ground Cooling

Since the ground is nearly always cooler than the air in the months when cooling is required, the more a house is in contact with the ground the cooler it will be. Build your house below grade or into the side of a hill to obtain easy ground contact. You can also partially bank (berm) the earth around your house, or even cover it.

High levels of comfort and serenity usually accompany well-designed underground housing. Just be careful to insulate the building in a way and to a degree that's appropriate to your region. In cold climates, insulate the walls well from the ground. In mild climates, use less insulation. In warm climates, no insulation is needed; the earth will keep the house cool. In humid climates, provide ventilation so that the surfaces in contact with the ground are kept dry.

Earth Pipes

Buildings built near natural caves have long used underground air masses to provide ventilation, needing only a little heating in most seasons. Earth-pipe systems for the same purpose have been designed, using pipe ranging in diameter from 4 to 12 inches. From 40 to 200 feet of pipe are buried 3 to 6 feet below the earth. Metal culverts and plastic and metal waste pipe are used. As house air is vented to the outside, either naturally or with a fan, outside air is drawn through the pipes and then into the house. During the winter, the air is warmed. During the summer, the air is cooled.

In humid climates, moisture condenses out of the air and onto the surfaces of the tubes. The pipes are sloped slightly outward, away from the house, to carry away the moisture left behind by the humid air.

A major drawback is that the earth surrounding such pipes, in many cases, is warmed (or cooled) too quickly to the temperature of the incoming summer (or winter) air, causing the pipes to lose their intended usefulness.

House Design Based on Climate

"Will passive solar heating and cooling work where I live?" Simply put, the answer is "Yes!" After all, the sun shines everywhere. However, the question is not whether it will work, but how. What is the best way for you to use the many applications we have described in this book? The most important factor influencing the answer to this question is climate.

A Tradition of Regional Architecture

The amount of time the sun shines differs from one part of country to another. Temperature variations are even greater. Wind conditions vary from calm to continuously windy. Some areas are arid, while others are humid.

It really wasn't so long ago that we used rather primitive methods of heating and cooling our homes: fireplaces in the winter and windows in the summer. We had limited access to materials to improve the thermal performance of our houses. We had to rely on local building materials such as logs, stone, adobe or sod. Glass was scarce, window screen non-existent, and sawdust substituted for insulation.

Combined with the cultural diversity of the immigrating settlers, the forces of climate and cheap available building materials resulted in a rich variety of architecture from one region to another.

94

For example, the harsh winters and abundant forests of the North Country combined to give us the traditional wood "saltbox" house. Its compact floor design is oriented to the south, and covered by a long north roof to shield it from the cold winter winds.

In the Great Plains, sod substituted for wood, and underground shelters offered protection from harsh westerly winter winds.

The oppressive humidity of the mid-Atlantic and Gulf states gave us narrow floor plans, floor-to-ceiling windows and doors, high ceilings and broad overhangs for enhancing summer ventilation.

In the Southwest, Spanish settlers used thick adobe walls and shaded courtyards with fountains to keep interiors cool during the summer and warm during the winter.

Simplified heating and cooling technology developed more quickly than improved materials and techniques for upgrading the thermal performance of houses, in part because of abundant and cheap energy. The result is that large central heating and cooling systems run by cheap energy compensate for climatically inappropriate house designs.

For example, torrid summer solar gains through large west-facing windows in the arid Southwest have become as commonplace as horrendous heat loss through large, north-facing single-pane windows in the brutally cold North.

Building materials are easily transported from one part of the country to another, at great energy expense. Petroleum-based plastics imitate adobe in the north and window shutters in the south.

Monotonous-looking subdivisions have become Anywhere, USA.

The era of unlimited cheap energy has passed. We again have the opportunity to design houses that work with the climate and not against it. To make best use of this opportunity, we must understand the wide variety of energy conservation and passive solar heating and cooling applications, so as to select a suitable combination for a particular climate. In doing so, we will obtain the highest possible comfort at the lowest possible expenditure for materials and energy.

Revival of Climate-Based Architecture

Climate, then, should have a major effect on your selection of energy conservation options and passive solar heating and cooling features, and, in turn, the architectural design of your house. The relationship between these factors and climate is discussed here to assist you with your choices. You will find that there are enough possibilities to offer the potential of near-zero heating and cooling bills nearly everywhere in the country, while still satisfying your other needs for a home.

For simplicity, we can categorize climate into three types: winter-dominated climates, summer-dominated climates, and climates that have a relatively balanced mix of winter and summer.

No doubt you know which climate type you live in without having to look at a map of the country divided into the three zones. Besides, hills, valleys, lakes and rivers can cause your micro-climate to differ significantly from what a map would tell you.

Winter-dominated Climates

Northern New England, New York, the upper Midwest, and most of the Rocky Mountain highlands are typical of cold, winter-dominated climates. Make your house compact in shape and consider earth-sheltering if your soil and terrain is not too rough. Pay attention to construction details for minimizing air leakage. Consider an air-to-air heat exchanger that introduces fresh air preheated by exhaust air. Use 10 to 15 inches of insulation everywhere. Equip entrances with airlock vestibules.

Locate most windows as close to due south as possible, but certainly away from the cold winds which usually prevail from north and west. If summers are mild, moderate-size east- and west-facing windows will rarely create overheating problems.

Avoid east- and west-facing glass where summers get hot. Alternatively, shade this glass well and, especially in such humid climates as New York, Pennsylvania, and the upper Midwest, use them to admit cooling breezes.

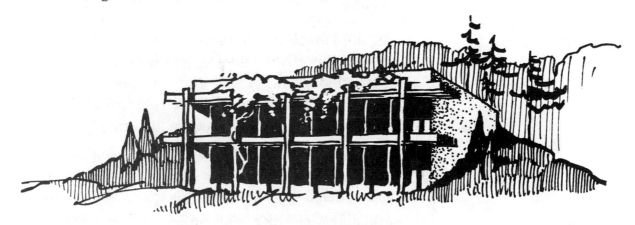

Roof overhangs can shade south glass and help trap gentle air currents for additional ventilation. Shade trees are particularly effective as "air conditioners" in winter-dominated climates that also have hot summers.

Many winter-dominated climates, such as much of the Rockies, experience mild summers, and cooling need not be a major consideration when designing a house to minimize winter fuel bills. Some parts of the Rockies, however, are hot during the day and cool at night. Therefore, thermal mass, important in this region in the winter for storing plentiful sunlight, cool down in the evening and absorb heat to keep houses comfortable during the day.

Supplementary evaporative cooling from pools, fountains, or "swamp" coolers may be needed in climates with hot, arid summers, such as the Great Plains and the plateaus of the Rockies.

In winter climates with abundant sunshine, such as Colorado and the Southwest, passive systems can be sized large enough to economically provide more than 75 percent of the heat. The greater the percentage of heat supplied by the sun, the more thermal mass is required for storing the excess heat for successive cloudy days. Thus, the most effective systems are solar windows in combination with concrete, stone, brick or adobe construction and solar walls. The same mass keeps these houses cool in the summer.

In most of the rest of the country the sun shines an average amount. Therefore, the choice of passive systems should be based on factors other than the availability of sun. For example, a solar room may be a good choice if additional moisture is desired in the house, but it may be a poor choice in a humid climate. On the other hand, a sun room in a humid climate can be converted to a screened summer porch.

The thermal mass of large solar window systems and solar walls is not as appropriate in winter-dominated climates that are also mild. In the Pacific Northwest, for example, windows and solar rooms with wood-framed houses will suffice without thermal mass. The windows can be used for summer ventilation.

Because passive systems lose more heat in cold weather than in mild, windows should have low-e or double glazing, and triple glazing in really cold climates, 9,000 degree days or more.

Solar chimneys and walls should be double-glazed in really cold climates. They are relatively more cost-effective than other passive systems in colder climates. Solar rooms produce less heat and cost more in cold climates. There may be other reasons for building them, however.

The extra layers of glazing or special coatings increase the cost of a cold-climate passive system. And since conservation saves more energy in cold climates than in mild ones, conservation can be carried to the maximum, and the passive system can be limited in size to 10 or 20 percent of the floor area.

Active solar systems, too, can be considered seriously because long heating seasons increase their cost effectiveness. Solar water heating, of course, should be used regardless of climate. And be sure to slope your roof southward for future conversion to solar electric cells or to additional solar heating.

Summer-dominated Climates

At the opposite end of the thermal spectrum are climates dominated by summer. Some parts of the country, such as Hawaii, Southern California and Florida, have no winter. Light, open construction permits houses to capture all available air movement. With proper shading, these houses should require no mechanical air conditioning other than occasional fans in humid climates and pools or fountains in arid ones.

In summer-dominated climates, winters (if they exist at all) are usually quite mild. People tend to design houses that can be buttoned up easily when temperatures drop. A design dilemma then results, especially in humid climates such as in the Gulf states. It may not be possible to open a house widely enough to permit sufficient cooling from ventilation. However, the same house may not be tight enough for efficient conventional heating and cooling.

A partial solution to this dilemma is a house that is both well insulated and well shaded. Windows should be double-glazed. For removing the winter chill, one third to one half of the total window area should face south. Properly designed houses will require little if any other heat in most winter climates of less than 3,000 degree days per year.

The rest of the windows should be strategically placed to capture and enhance breezes. Locate large windows or vents as high as possible. Belvederes, turrets, gables, wind turbines and fans will add to the ventilating chimney effect. Place intake windows close to the floor and, if possible, facing the winds. High ceilings help restrict warm air to levels above people's heads. Avoid shade trees in humid climates since they tend to stifle breezes and their respiration adds moisture to air.

Make roofs light in color and vent all attic space well. Elevate houses slightly to assure good drainage and to minimize humidity. Carefully moisture-proof all floors and walls in direct contact with the ground but do not insulate unless cold, moist interior surfaces are considered undesirable.

If you use earth pipes for cooling, make sure they are moisture-proof to avoid further humidification of incoming air by the ground. This cooling method does not work in climates without winters. Even in climates with mild winters the pipes must be buried more than six feet deep to obtain sufficiently cool temperatures.

In arid climates, earth contact can provide pleasant relief from hot weather and does not have associated moisture problems, since humidity is often encouraged. Earth pipes can be perforated to increase humidification of the air by the ground.

Thermal mass is usually of little advantage in mild, humid climates. Solar windows and chimneys can accomplish most if not all of the heating. Solar chimneys can also be used to enhance summer ventilation.

Solar walls are less applicable unless summer nights are mild, in which case the mass of the wall can help keep the house cool during the day.

Solar rooms work best in humid climates if moisture respiration from plants is minimized. A good solar room in this climate is one which is easily converted to a screened porch during the summer.

In general, passive systems can be single-glazed. However, double-glazing of windows can help eliminate the need for a heating system or, in less mild winters, reduce the size of the system to a single space heater.

The design strategy is somewhat different in arid, summer-dominated climates such as Southern California and the deserts of the southwest. Conventional construction is of massive materials, such as adobe. Tall, narrow east- and west-facing windows are recessed for easy shading.

South walls of heavy materials need not be covered with glass or plastic in mild areas. They will provide adequate heat in the winter and can be shaded to reduce solar gains during the summer. East, west, and north walls can be insulated to reduce both winter heat loss and summer heat gain.

The addition of a second layer of glazing to windows can eliminate fuel bills in many arid, summer-dominated climates. Extra south windows, or a solar room, may be needed to achieve this.

Cooling bills can also be eliminated by taking further steps. Pools, fountains, and other evaporative cooling techniques are ideal, except where water is a scarce resource. Pool ponds and sprays work well, too. Radiational cooling is effective if night skies are clear, and solar roofs should be considered. Walls and roofs should be light in color to reflect summer heat.

A Balance Between Winter and Summer

In many parts of the country, neither winter nor summer dominate. Year 'round mild weather prevails in San Francisco and its environs, but in the southern Midwest both summer and winter are equally harsh. Southern New England is considered moderate.

First, the mild areas. These are the climates of greatest design freedom. Thick insulation, proper orientation, and a modest-sized passive system can all but eliminate heating bills.

Shading, thermal mass with night cooling, and ventilation can do most or all of the cooling. Radiational cooling using roof panels can also be used.

Moderate climates hold many of the same opportunities as mild climates; they offer numerous design choices. However, mistakes are not as forgiving, so more careful design is necessary.

Elimination of backup heating and cooling is more difficult, but possible.

The best approach is to find local houses 50 to 100 years old that are comfortable both summer and winter. For your own design, retain their thermal properties and embellish them further with energy conservation and passive heating and cooling measures. Chances are, you will have a very successful low-energy house.

Harsh climates may be either humid or arid. It is the high fuel bills that are typical of these climates that offer both the substantial incentive and the substantial rewards for using energy conservation and passive solar. Review the earlier parts of this chapter and combine heating and cooling measures that do not conflict with one another. Lots of insulation and multipaned windows, for example, rarely compete with other measures.

The Hard Choices of Cost

We are now prepared to determine the proper size of a passive system. "Size" almost always refers to square footage of all glazed surfaces facing the sun, within 30 degrees east or west of due south. Other dimensions, such as the thickness of a solar wall or the floor area of a greenhouse, are fairly standard and we have already covered them.

The size of a system is the net square footage of glazed surface after subtracting the framing and trim. A wall of solar windows 8 feet high and 20 feet long (160 square feet) may be as much as 40 percent framing and 60 percent glazing, leaving but 96 square feet of window to catch the sun. It is possible, but difficult, to minimize framing to 5 or 10 percent of the surface area.

There are many ways to size a passive system. Engineers and scientists have produced dozens of documents crammed full with charts, graphs and tables for determining the precise optimal size and performance of passive systems. This detailed information has its uses, but do not feel obligated to use it or to rely on it. Remember that Nature is forgiving and does not require nearly the precision that our computers tell us we do! Instead, review again some of the earlier pages that discuss size and you will have a good feel for the surface area you need.

How Large?

The first question you should answer is this: How large do you want the system to be? Believe it or not, answering this question is by far the most common, direct and useful method of determining size. Most people want their system to be as large as possible to save as much energy as possible. In general, this is an excellent approach to take, and you may want to determine how to cover the entire south side of your house with passive devices. Many people do, just as the size of the roof often determines the size of active solar heating collectors.

You need not go to this extent, of course. A superinsulated house can be 50 percent solar heated with only slightly larger than normal south windows. But if you want to reduce your fuel bills to an absolute minimum, large systems are generally in order. The trick is to design them right, so that you save as much energy as possible with the least expense, and so that you are pleased with the appearance and comfort of your home.

Generally speaking, an optimal size for a passive system is that which supplies the same portion of heat as the average percent of the time that the sun shines during the winter. For example, if your heating season is October through April, determine your local average winter sunshine from the U.S. monthly sunshine maps in Appendix 3. This percentage, say 55 percent for much of the country, is the portion of your heating load that can be supplied by a passive system in a well-insulated house with good assurance that the design can look beautiful, perform well and be cost effective.

This is not to say that you should not try for larger percentages to be supplied by solar. But the design gets trickier, so consider obtaining additional advice from a local expert, a workshop or more books. The bibliography at the back of this book can provide some helpful hints on where to look next.

Doing the Calculations

The next step is to determine your annual fuel consumption in gallons of oil, kilowatt hours (kwh) of electricity, or hundreds of cubic feet of gas (ccf). For an existing house, look at fuel bills. For new houses, calculate heat loss (get help or consult books—it's easy!). It's not unusual for an old, rambling Victorian house to use during a cold winter several thousand gallons of oil or kwhs of electricity, or several hundred thousand cubic feet of gas. But a compact, superinsulated house may use as little as one fifth this amount—even before the addition of passive devices.

Oil, Electricity, Gas

A gallon of oil would supply 135,000 to 140,000 Btus if your furnace burned at 100 percent efficiency. It doesn't. Most furnaces, in fact, burn at between 40 and 70 percent annual efficiency, supplying the house with between 56,000 and 100,000 Btus per gallon burned. A useful average to use is 70,000.

One kwh of electricity is equivalent to 3400 Btus. Most electric resistance heating is 100 percent efficient, although three to four times more energy is expended at a power plant than what finally enters your house as electricity. Thus, 20 kwhs (68,000 Btus) of electric resistance heating supplies the heat obtainable from one gallon of oil (70,000 Btus). Moreover, proper application of an electric heat pump can supply twice as much energy per kwh as resistance heating.

One cubic foot of gas contains 1,000 Btus of heat. One hundred cubic feet (1 ccf) contains 100,000 Btus. At 60 percent annual furnace efficiency, each ccf of gas supplies 60,000 Btus, close to that of one gallon of oil or 20 kwh of electric resistance heating.

The next step is to determine the energy output of the system. This seems like the hardest step, requiring the charts and graphs we mentioned earlier. However, it need not be difficult. In fact, it can be relatively easy.

Each square foot of a design that is appropriate for the climate—for example, a design that has the proper number of glazing layers—will supply the heat obtainable from roughly one gallon of oil (60,000 Btus). In cloudy climates, such as Buffalo, New York, the amount can be half that. In the sunny Southwest the output can double.

Thus, a house with an annual heat loss of 500 gallons in a climate of average (50%) sunshine will require 250 square feet of passive system to halve its fuel bill.

When selecting among the various passive applications, review the previous chapter for their appropriateness to your climate. In general, start with windows as the basic system and add other systems as needed or desired.

In the example just above, the home may be designed for 100 square feet of south windows. They will replace 100 gallons of oil or 20% of the total 500-gallon load. As discussed earlier, no special effort needs to be taken to add thermal mass, since conventional construction usually provides adequate storage for solar windows that supply 20 percent or less of the heat.

The next 150 square feet, supplying 30 percent of the load, might, for example, consist of a greenhouse, a solar wall, or a combination of the two. Remember to take into account the trim and framing when determining square footage.

Conserve First

Don't be fooled by the glamour of solar energy.
Energy conservation, perhaps less glamorous, can go
a long way toward saving energy. We can
economically superinsulate homes so that they use as
little as 20 percent of the energy that conventional
homes use, without significant use of solar energy.

In moderately cold climates, walls and roofs should
have R-values of at least 25 and 40, respectively. Use
low-e or triple glazed windows, and caulk and
weatherstrip. In cold climates, increase these R-values
of the walls and roofs by 50 percent. If you do
anything less than these measures you should restrict
your expenditures on solar. If you meet these
measures, go all out with passive solar.

From a practical point of view, energy conservation
measures are necessary in most of the U.S. climate to
keep the required solar heat collection area small
enough so that it can be easily and economically
added to the average house. Properly designed solar
glazing totalling 20 to 40 percent of the floor area of
your house (depending on the climate), will provide
50 to 80 percent of the heat. This is for a well-
insulated house with adequate thermal mass for heat
storage. (Superinsulated houses need less southern
exposure to achieve this.) Thus, the collection glazing
can cover much of the south-facing walls (or roof).
Poorly insulated houses cannot normally obtain a
large percentage of heat from the sun except in mild
climates.

In cold, cloudy climates, energy conservation
measures are usually more cost effective than are
solar heating devices, at least until heating loads of a

house are reduced by 40 to 60 percent compared with previous heating bills. But the house is still a "conventional" one compared with a superinsulated house. In mild, sunny climates, solar heating and cooling may be a more cost-effective means of reducing fuel bills than energy conservation, since the climate permits passive systems to have fewer layers of glazing.

Keep It Simple

Passive system costs can vary enormously, depending on size, design, materials, construction methods and passive system type. Keep the sizes manageable. For example, a 15-foot-high wall of glass is costlier to build than an 8-foot-high wall. Don't make the design complicated. Keep it simple! An occasional fan to help move air from one part of the house to another—for example, from the south side to the north or from upstairs to downstairs—can keep floor plans from becoming contorted and passive systems from becoming contrived.

Use basic, easy-to-handle materials, preferably locally produced. Build the systems yourself, if possible. This can save 40 to 60 percent of the cost. Contractor-built systems are more cost-effective on new housing than on existing housing. However, additions and major remodelings can be cost-effective contractor-built systems for existing houses.

Trade-offs

Which of the various passive system types you choose also affects cost. Solar windows cost nothing if what would have been a north window is instead placed facing south. Solar windows can also be inexpensive if you are willing to have some of them fixed in place, nonoperable. Fixed glass can be one third the cost of operable windows.

If you intend to build of heavy materials regardless of using solar, your thermal mass comes free. On the other hand, concrete floor slabs or brick partitions added to conventional wood-framed construction can increase the cost.

In mild climates, windows need have only two layers. In severe climates, low-e or triple glazing are usually necessary (and cost effective!). Shades or awnings may add to the cost in southern climates.

If you are able to convert your south wall easily to a solar chimney by simply adding a layer of glazing, the materials will cost but a few dollars per square foot. If you buy a solar collector and hire someone to install it, the final cost will be much more. Either way, if the collectors are large enough to provide only 15 to 20 percent of the heat, no new thermal mass will be needed.

The cost of making your south wall into a solar chimney, then, will be primarily that of just the collector or glazing. When more than 20 percent heating is obtained, however, thermal mass must be added to accommodate the heat. This makes solar chimneys a whole new ball game, and of course increases cost.

The cost of building a solar wall is less when the south wall would have been built of concrete block, brick, stone or adobe anyway. Keep in mind that stronger foundations are needed to support the weight, and that extra floor area may be needed to accommodate the thicker than normal wall. In mild climates only single glazing may be needed, while in cold climates as many as three layers of glass or movable insulation may be necessary.

Conventional heating and cooling systems can be greatly reduced in size and complexity, or eliminated entirely, from many energy-conserving, passive solar homes throughout the country, saving several thousands of dollars in first cost and hundreds of dollars in maintenance costs. Small space heaters or wood-burning stoves often suffice in the winter, and fans or modest-sized window air conditioners or evaporative coolers do the job in the summer.

A single-glazed solar room can cost only a few dollars if the frame is built of used lumber and covered with a thin sheet of plastic. Commercially sold greenhouses, however, can cost thousands of dollars. Solar rooms reduce the cost of other building components. For example, when protected by a solar room, walls that would otherwise be exposed to the weather can be simplified and reduced in cost.

Whatever system or combination of systems you choose, shop around. Seek advice and look for the best prices. Pay attention to detail and use passive systems to enhance the beauty of your house.

The Real Reward

Will you be satisfied with the fuel savings resulting from your expenditures for passive systems? The easy answer is "Yes!" However, expectations and needs are different. For example, some people want passive to pay for itself in five years. Others can afford to be more farsighted and view conservation and solar design as investments in the long-term future.

Another way to view many passive systems is as an integral part of your house, increasing in value along with the whole house. Thus, when the value of a house inflates over say, five years, so will its sunroom or attached greenhouse! If the house is then sold, the owner gets back the investment in the solar room—plus five years of lower energy bills.

In this sense, solar energy is, indeed, free. But passive solar is also convincing on the basis of dollars spent for the energy saved. For the many solar and conservation measures recommended in this book, a sizeable expenditure upfront may be necessary, depending on climate, the size of the house, and which conservation and solar features are selected. But the result will be fuel bills reduced by 70 to 90 percent.

Afterword

You have now reached the end of the book and are at least seven percent more knowledgeable than you were when you started it. Now it's time to put all your new-found knowledge to work. The things you've learned are of great value, for they can affect not only your pocketbook, your comfort and your way of life, but global politics, environmental quality and the value of the dollar as well.

We are in a period of fundamental transition, from the fossil-fuel age to the solar age. Standing close to that transition, it's sometimes hard to sense the magnitude of the changes now taking place. But each day brings a bit more perspective to our view of the transformation. Each day makes solar energy systems look better and conventional sources of energy look worse.

Each step we take in the direction of a direct solar-energy-based civilization will have untold benefits to us, to our children, and to all the generations to come, not to mention the benefits to all the other creatures of the world, whose fate is so inextricably tied to ours.

Construction Details

This section features construction details from some of the more typical examples of the passive systems we have described. This information should give designers, builders, and homeowners a much better knowledge of how each system is designed and built, with a focus on those details needing special attention.

The details shown here are from four designs developed in two federally sponsored demonstration project to promote solar design, research and construction. They include a solar room, a solar window and a solar chimney from Project SUEDE, and a solar wall from the Brookhaven House. The goal of the Brookhaven project was to develop an attractive, energy-conserving, single-family home of conventional design, using thermal storage materials in combination with heavy insulation and passive solar systems to cut heating costs significantly without reducing comfort.

Since full sets of construction drawings are not included here, you may wish to seek professional assistance before actually building your own system.

It is good practice to have all dimensions, quantities, and specifications reviewed by a competent local architect, engineer and/or building official prior to construction to assure compliance with individual requirements and local codes and conditions.

The drawings here were prepared by and adapted by the authors from designs for the Brookhaven and SUEDE projects. As neither the authors, publisher, nor any of the original SUEDE and Brookhaven project participants have any control over the final use of these revised drawings, all warranties, expressed or implied, for the usefulness of these drawings and all liabilities that may result from the use of these drawings are voided by their use in construction.

Project SUEDE, "Solar Utilization, Economic Development and Employment," was part of a nationwide effort to train solar installers and to build solar applications into existing houses. A major goal in Project SUEDE was to demonstrate that solar designs can be simple, can be built at reasonable costs from readily available building materials, and can be attractive and work well.

Sponsored by the U.S. Community Services Administration, Department of Energy and Department of Labor, SUEDE was carried out in New England by a four-member consortium: the Center for Ecological Technology in Pittsfield, Massachusetts; the Cooperative Extension Service of the University of Massachusetts in Amherst, Massachusetts; Southern New Hampshire Services in Manchester, New Hampshire; and Total Environmental Action Foundation in Harrisville, New Hampshire. Together, these groups trained 30 installers who built one of three types of low-cost solar systems onto nearly 100 New England homes.

In the examples of the New England SUEDE systems illustrated here, the added solar windows and the thermosiphoning air panel retrofit were each constructed by the Center for Ecological Technology. The attached greenhouse was constructed by Southern New Hampshire Services. The designs for New England SUEDE were developed by Total Environmental Action (TEA), Inc.

The Brookhaven House is the result of a research and design effort carried out by TEA, Inc., under contract to Brookhaven National Laboratory, and built at the Lab site on Long Island as a demonstration house to be monitored for its performance. The work was sponsored by the Building Division of the Office of Buildings and Community Systems, Office of the Assistant Secretary of Conservation, and Solar Applications, U.S. Department of Energy.

Solar Rooms

This solar greenhouse uses stock-size insulated glass patio door units as the solar aperture. These units are field mounted in the wood-framed structure which rests on an added foundation wall of poured concrete or block, and which is attached to the existing house wall by 2 x 4 braces and a 2 x 4 ledger strip bolted to the wall. The side wall can be either clapboard or other siding to match the house. In this design, the two-inch extruded polystyrene foundation insulation is located on the inside of the foundation wall to make a weatherproof exterior with no additional finishing required. An optional roll-down insulating curtain is included at the sloped glazing. (Construction details, New England SUEDE.)

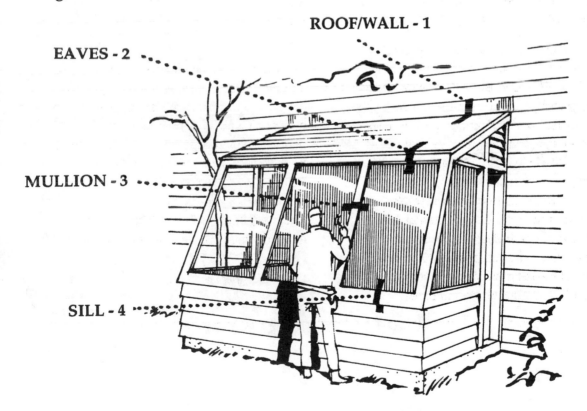

ROOF/WALL - 1

EAVES - 2

MULLION - 3

SILL - 4

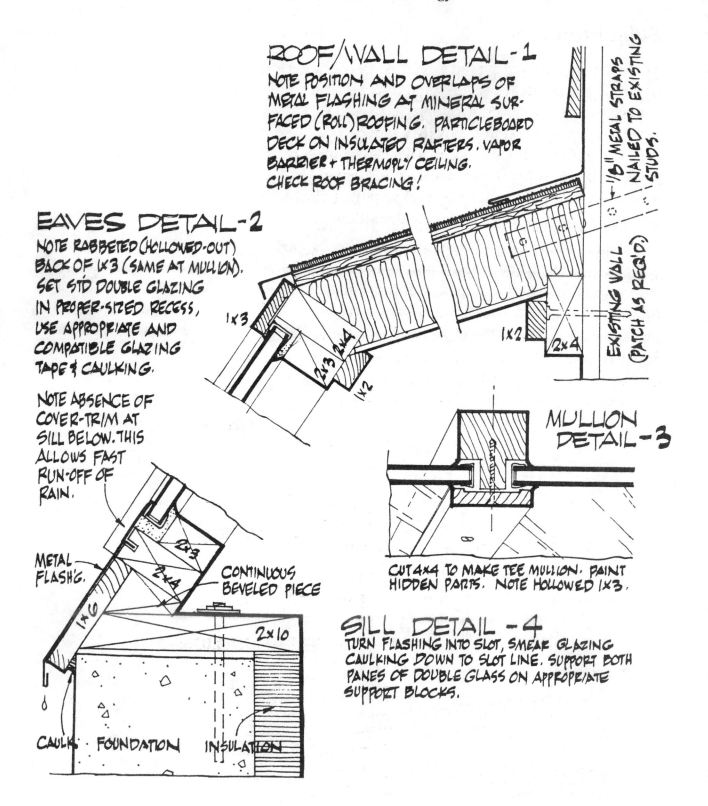

ROOF/WALL DETAIL-1
NOTE POSITION AND OVERLAPS OF
METAL FLASHING AT MINERAL SUR-
FACED (ROLL) ROOFING. PARTICLEBOARD
DECK ON INSULATED RAFTERS. VAPOR
BARRIER + THERMOPLY CEILING.
CHECK ROOF BRACING!

EAVES DETAIL-2
NOTE RABBETED (HOLLOWED-OUT)
BACK OF 1x3 (SAME AT MULLION).
SET STD DOUBLE GLAZING
IN PROPER-SIZED RECESS,
USE APPROPRIATE AND
COMPATIBLE GLAZING
TAPE & CAULKING.

NOTE ABSENCE OF
COVER-TRIM AT
SILL BELOW. THIS
ALLOWS FAST
RUN-OFF OF
RAIN.

METAL
FLASH'G.

CONTINUOUS
BEVELED PIECE

CAULK   FOUNDATION   INSULATION

1x3

2x4

1x2

2x4

EXISTING WALL
(PATCH AS REQ'D)

1/8" METAL STRAPS
NAILED TO EXISTING
STUDS.

1x3

2x4

1x6

2x10

MULLION
DETAIL-3

CUT 4x4 TO MAKE TEE MULLION. PAINT
HIDDEN PARTS. NOTE HOLLOWED 1x3.

SILL DETAIL-4
TURN FLASHING INTO SLOT, SMEAR GLAZING
CAULKING DOWN TO SLOT LINE. SUPPORT BOTH
PANES OF DOUBLE GLASS ON APPROPRIATE
SUPPORT BLOCKS.

Solar Windows

These details were developed for a low-cost addition of direct gain south glazing in standard 2 x 4 stud wall construction. A section of the south wall is removed and new framing added as shown to prepare for the addition of standard-sized insulated glass units. These fixed units are installed using standard glazing techniques, including setting blocks, glazing tape and weep holes for condensation.

The rough framing is finished with trim pieces and glazing stops. Note that cutting into the framing of a stud wall house can be a major structural alteration to he house and should only be undertaken after professional verification that the new structure is adequate and that existing floor and roof loads can be carried safely during and after the renovation project. (Construction details, New England SUEDE.)

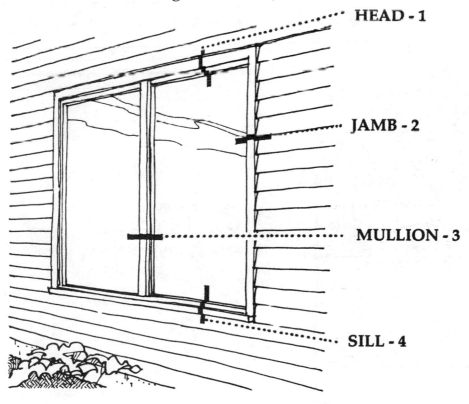

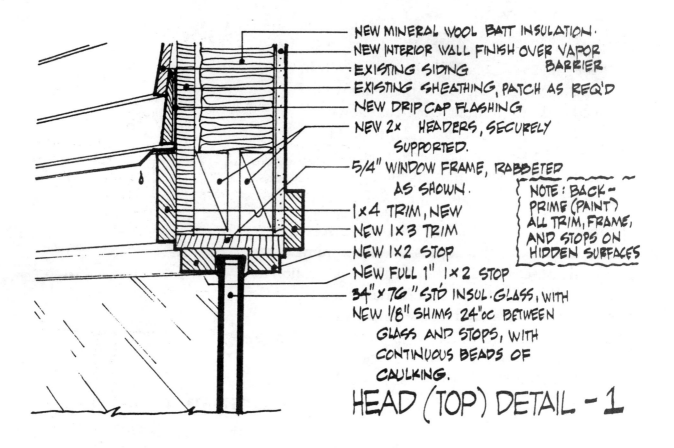

NEW MINERAL WOOL BATT INSULATION.
NEW INTERIOR WALL FINISH OVER VAPOR
EXISTING SIDING   BARRIER
EXISTING SHEATHING, PATCH AS REQ'D
NEW DRIP CAP FLASHING
NEW 2x HEADERS, SECURELY
    SUPPORTED.
5/4" WINDOW FRAME, RABBETED
    AS SHOWN.
1x4 TRIM, NEW
NEW 1x3 TRIM
NEW 1x2 STOP
NEW FULL 1" 1x2 STOP
34" x 76" STD INSUL. GLASS, WITH
NEW 1/8" SHIMS 24"OC BETWEEN
    GLASS AND STOPS, WITH
    CONTINUOUS BEADS OF
    CAULKING.

NOTE: BACK-
PRIME (PAINT)
ALL TRIM, FRAME,
AND STOPS ON
HIDDEN SURFACES

HEAD (TOP) DETAIL - 1

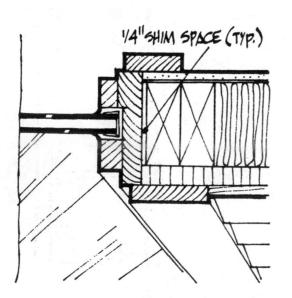

1/4" SHIM SPACE (TYP.)

MEMBERS SHOWN
AT LEFT ARE THE SAME
AS THOSE NOTED ON THE DETAIL ABOVE.
NOTE ABSENCE OF FLASHING.
CAULK BETWEEN SIDING AND 1x4.

JAMB (SIDE) DETAIL- 2

AGAIN, THE DETAIL IS SIMILAR
TO THE HEAD DETAIL
BUT NOTICE THAT THE
5/4" WINDOW FRAME
IS 3-1/2" TO MATCH THE 2×4s,
AND THE COVER TRIM,
INSIDE AND OUT, IS
4 1/2" WIDE

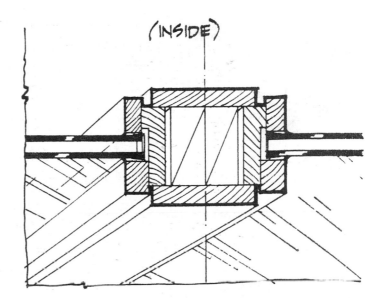

(INSIDE)

MULLION (POST) - 3

SILL DETAIL - 4

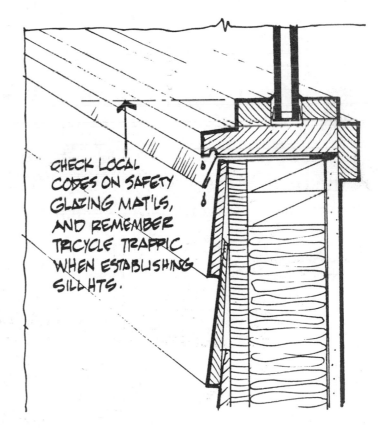

CHECK LOCAL
CODES ON SAFETY
GLAZING MAT'LS,
AND REMEMBER
TRICYCLE TRAFFIC
WHEN ESTABLISHING
SILL HTS.

NOTE THAT THE
WINDOW FRAME AT THE SILL
IS MADE OF 2X MATERIAL
WITH TWO RABBETED STEPS
ON TOP AND ONE
DRIP SLOT ON THE
BOTTOM.
CAULK THE UNDERSILL FLASHING
NEAR THE INDOOR SIDE.

Solar Chimneys

This retrofit passive space-heating device, called a thermosiphoning air panel (TAP), uses the existing house wall as the major structural element. The exterior finish is removed, new Thermoply™ structural sheathing added over the existing wall, and wood framing added to support the field-installed insulated glass units.

The system shown uses three patio door replacement units as the aperture, creating three areas of absorber plate, each of which requires a high and a low vent through the house wall to allow the thermosiphoning action to occur. The weight of the added glazing is carried by brackets at the base of the panel to a continuous ledge strip bolted to the house wall. After flashing is added, the exterior siding materials are patched around the unit to complete the installation. (Construction details, New England SUEDE.)

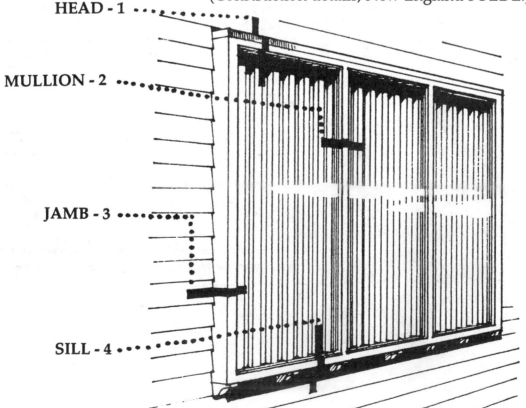

HEAD - 1

MULLION - 2

JAMB - 3

SILL - 4

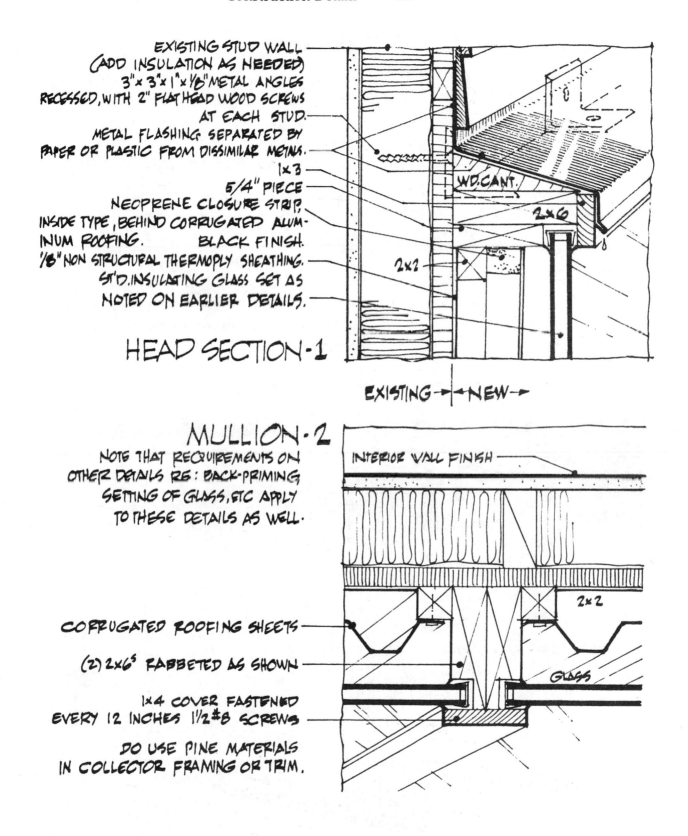

EXISTING STUD WALL
(ADD INSULATION AS NEEDED)
3"x 3"x 1"x 1/8" METAL ANGLES
RECESSED, WITH 2" FLAT HEAD WOOD SCREWS
AT EACH STUD.
METAL FLASHING SEPARATED BY
PAPER OR PLASTIC FROM DISSIMILAR METALS.
1x3
5/4" PIECE
NEOPRENE CLOSURE STRIP,
INSIDE TYPE, BEHIND CORRUGATED ALUM-
INUM ROOFING.   BLACK FINISH.
1/8" NON STRUCTURAL THERMOPLY SHEATHING.
ST'D. INSULATING GLASS SET AS
NOTED ON EARLIER DETAILS.

WD. CANT.

2x6

2x2

HEAD SECTION · 1

EXISTING → ← NEW →

MULLION · 2

NOTE THAT REQUIREMENTS ON
OTHER DETAILS RE: BACK-PRIMING
SETTING OF GLASS, ETC APPLY
TO THESE DETAILS AS WELL.

INTERIOR WALL FINISH

2x2

GLASS

CORRUGATED ROOFING SHEETS

(2) 2x6's RABBETED AS SHOWN

1x4 COVER FASTENED
EVERY 12 INCHES 1 1/2 #8 SCREWS

DO USE PINE MATERIALS
IN COLLECTOR FRAMING OR TRIM.

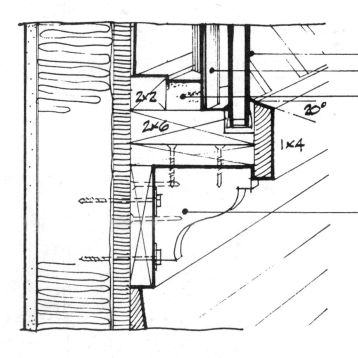

INTERIOR WALL FINISH.

CAULK

SIDING

2×2

5/4 PC.

2×6

GLASS

1×3

SELECT 2×6 MEMBERS CAREFULLY!

LEAVE ¼" GAP AT SIDING FOR CAULKING;
USE MAT'L COMPATIBLE WITH FINISH ON
WOOD.

NOTE THAT VERTICAL PIECES ARE NOT
FASTENED TO EXISTING WALL; HORIZON-
TAL MEMBERS ARE.

JAMB DETAIL - 3

SILL DETAIL - 4

GLASS.
CORRUG. METAL AND CLOSURE STRIP.
NEOPRENE CLOSURE STRIP

2×2

2×6

20°

1×4

SUPPORT BRACE: 2 CONTINUOUS 1×6 OR
5/4"×6" BOARDS WITH 2" BRACKETS
EVERY FOOT. LONG FLATHEAD SCREWS
FASTEN BRACE TO BRACKETS.

CONNECT BRACE TO WALL WITH
(2) 3" LAG SCREWS @ EA. STUD

Solar Walls

The construction details shown here are from the triple-glazed storage wall located next to a large sunspace and serving as the structural south wall of the dining room. This storage wall also contains a set of windows for direct gain, natural lighting and a view from inside.

This glazed thermal storage wall is comprised of glazing frame members milled from cedar 4 x 4s bolted to an eight-inch thick structural brick wall. The bricks are dense paving bricks—a dark umber color on the outside, standard terra cotta color on the inside—and are laid up with all cavities filled with mortar.

The triple glazed panels, designed for use in the northeast, reduce heat losses to the outside from the warm wall. Standard operable triple-glazed casement windows are incorporated into the wall to provide direct gain heating, light, views and ventilation. Double glazing is suitable for us in milder climates. (Construction details, the Brookhaven House.)

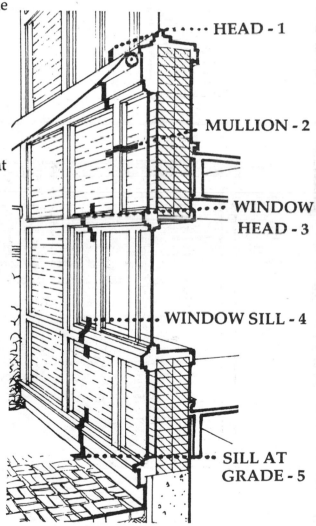

HEAD - 1

MULLION - 2

WINDOW HEAD - 3

WINDOW SILL - 4

SILL AT GRADE - 5

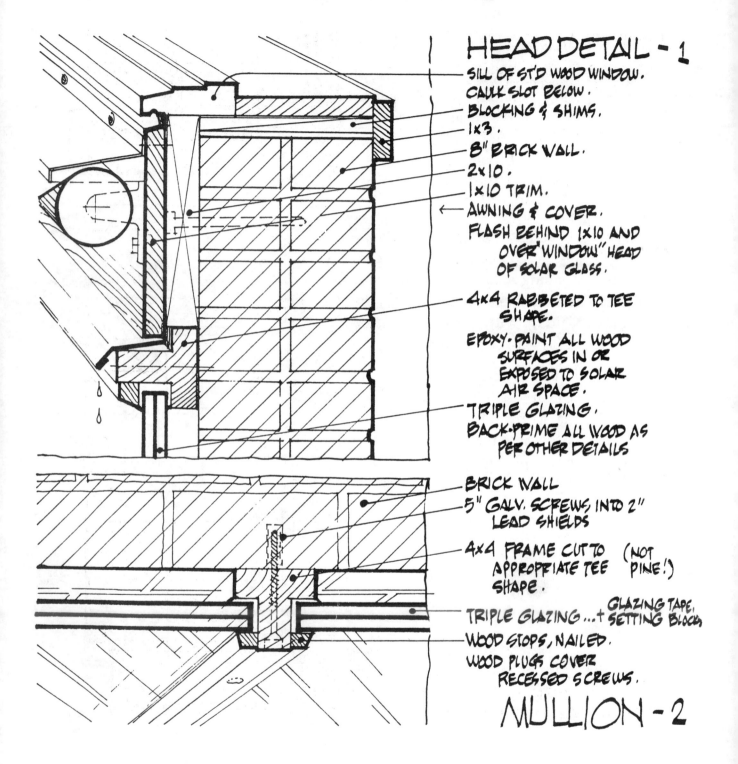

HEAD DETAIL - 1

SILL OF ST'D WOOD WINDOW.
CAULK SLOT BELOW.
BLOCKING & SHIMS.
1x3 .
8" BRICK WALL.
2x10 .
1x10 TRIM.
← AWNING & COVER.
FLASH BEHIND 1x10 AND
    OVER "WINDOW" HEAD
    OF SOLAR GLASS.

4x4 RABBETED TO TEE
    SHAPE.
EPOXY. PAINT ALL WOOD
    SURFACES IN OR
    EXPOSED TO SOLAR
    AIR SPACE.
TRIPLE GLAZING .
BACK-PRIME ALL WOOD AS
    PER OTHER DETAILS

BRICK WALL
5" GALV. SCREWS, INTO 2"
    LEAD SHIELDS

4x4 FRAME CUT TO   (NOT
    APPROPRIATE TEE   PINE!)
    SHAPE.
                    GLAZING TAPE,
TRIPLE GLAZING ...+ SETTING BLOCKS
WOOD STOPS, NAILED.
WOOD PLUGS COVER
    RECESSED SCREWS.

MULLION - 2

WINDOW HEAD - 3

BACK-TO-BACK STEEL ANGLES MUST BE SIZED TO CARRY TOTAL MASONRY WALL LOAD.

FLASHING EXTENDS UP BEHIND TEE.

GET GOOD ADVICE ON FASTENING WOOD TO STEEL AND BRICK.

WINDOW SILL - 4

METAL FLASHING IN ⌐ SHAPE PROTECTS 4x4 TEE.

STANDARD WDW. SILL IS SIMILAR TO DETAIL ON OPPOSITE PAGE.

TRIPLE GLAZING.

SILL AT GRADE - 5

TRIPLE GLAZING IN 4x4 TEE AS BEFORE.

SILL PIECE WITH DRIP.

2x4

2x4

EXTERIOR GRADE.
1/2" CEMENT PLASTER ON CHICKEN WIRE PROTECTS 2" POLYSTYRENE BOARD INSULATION.

CONCR. FOUNDATION

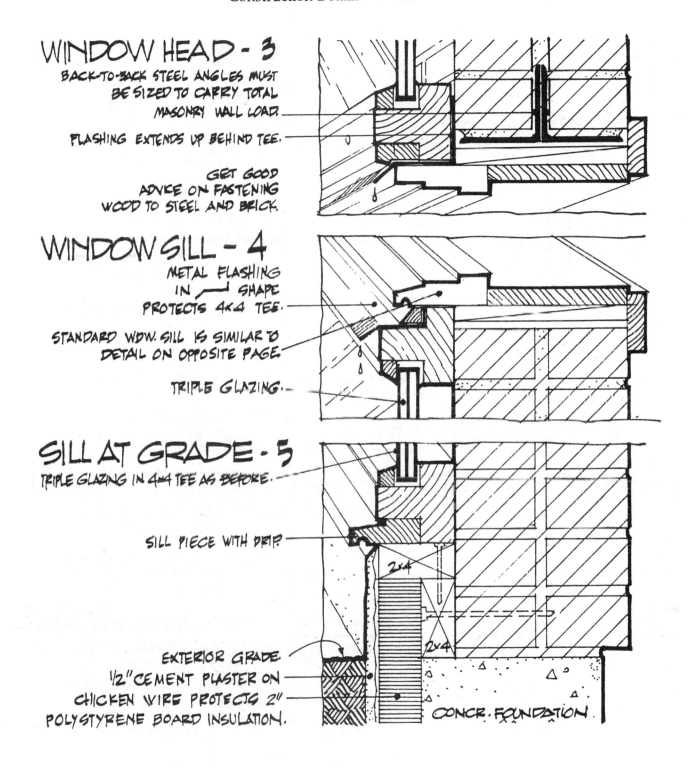

Superinsulated Wall

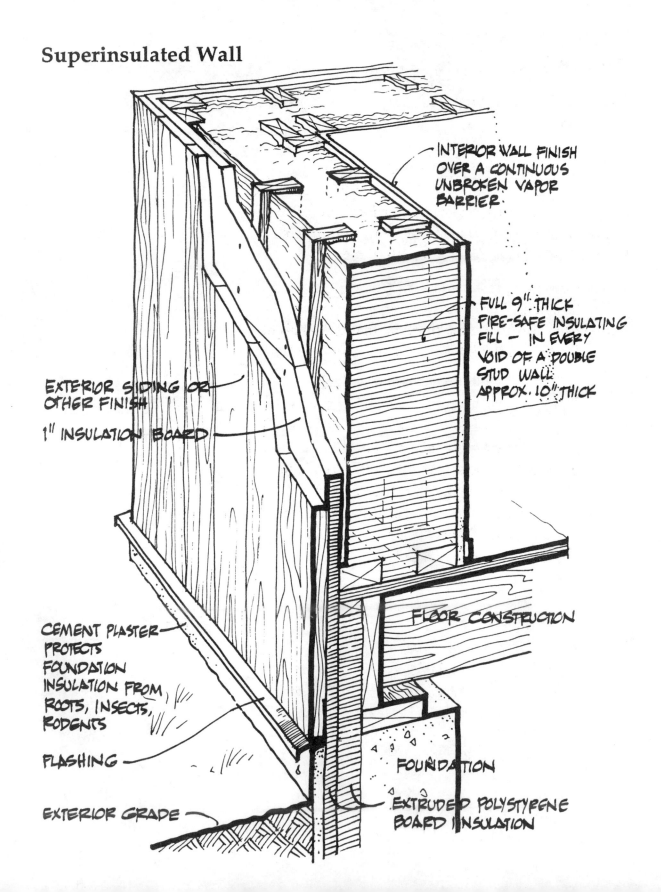

INTERIOR WALL FINISH OVER A CONTINUOUS UNBROKEN VAPOR BARRIER.

FULL 9" THICK FIRE-SAFE INSULATING FILL — IN EVERY VOID OF A DOUBLE STUD WALL APPROX. 10" THICK

EXTERIOR SIDING OR OTHER FINISH

1" INSULATION BOARD

FLOOR CONSTRUCTION

CEMENT PLASTER PROTECTS FOUNDATION INSULATION FROM ROOTS, INSECTS, RODENTS

FLASHING

FOUNDATION

EXTERIOR GRADE

EXTRUDED POLYSTYRENE BOARD INSULATION

Appendix 1

Sun Path Diagrams

Sun path diagrams are representations on a flat surface of the sun's path across the sky. They are used to easily and quickly determine the location of the sun at any time of the day and at any time of the year. Each latitude has its own sun path diagrams.

The horizon is represented as the outer circle, with you in its center. The concentric circles represent the angle of the sun above the horizon, that is, its height in the sky. The radial lines represent its angle relative to due south.

The paths of the sun on the 21st day of each month are the elliptical curves. Roman numerals label the curves for the appropriate months. For example, curve III (March) is the same as curve IX (September). The vertical curves represent the time of day. Morning is on the right (east) side of the diagrams and after noon on the left (west).

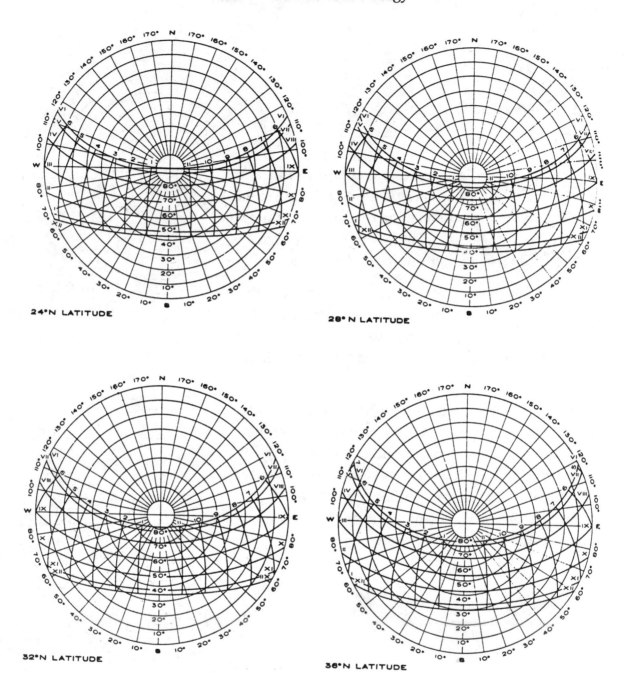

24°N LATITUDE

28°N LATITUDE

32°N LATITUDE

36°N LATITUDE

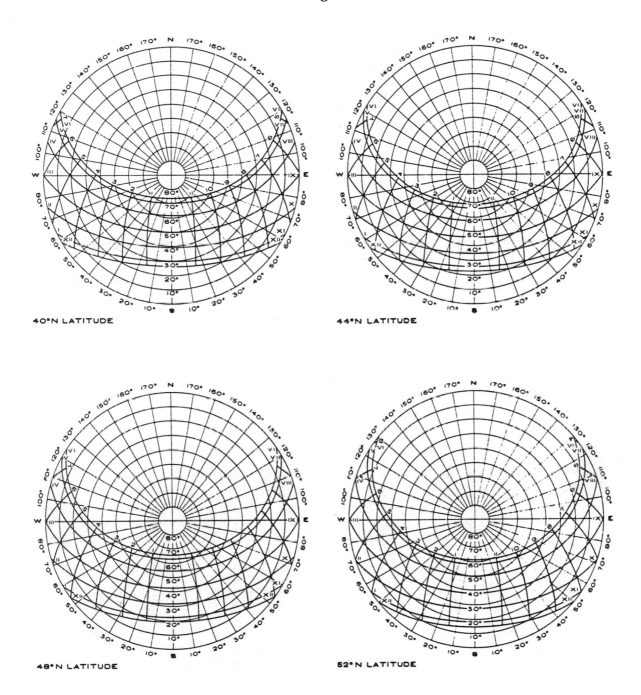

40°N LATITUDE

44°N LATITUDE

48°N LATITUDE

52°N LATITUDE

Appendix 2

Solar Radiation on South Walls on Sunny Days

Date	Solar Time		Solar Radiation, Btus per hour per square foot					
	AM	PM	Latitude (North)					
			24	32	40	48	56	64
Jan 21	7	5	31	1	0	0	0	0
	8	4	127	115	84	22	0	0
	9	3	176	181	171	139	60	0
	10	2	207	221	223	206	153	20
	11	1	226	245	253	243	201	81
	12		232	253	263	255	217	103
	Daily Totals		1766	1779	1726	1478	1044	304
Feb 21	7	5	46	38	22	4	0	0
	8	4	102	108	107	96	69	19
	9	3	141	158	167	167	151	107
	10	2	168	193	210	217	208	173
	11	1	185	214	236	247	243	213
	12		191	222	245	259	255	226
	Daily Totals		1476	1644	1730	1720	1598	1252
March 31	7	5	27	32	35	35	32	25
	8	4	64	78	89	96	97	89
	9	3	95	119	138	152	154	153
	10	2	120	150	176	195	205	203
	11	1	135	170	200	223	236	235
	12		140	177	208	232	246	246
	Daily Totals		1022	1276	1484	1632	1700	1656

* Courtesy ASHRAE, *Handbook of Fundamentals.*

Date	Solar Time		Solar Radiation, Btus per hour per square foot					
	AM	PM	Latitude (North)					
			24	32	40	48	56	64
Apr 21	6	6	2	3	4	5	6	6
	7	5	10	10	12	21	29	37
	8	4	16	35	53	69	82	91
	9	3	41	68	93	115	133	145
	10	2	61	95	126	152	174	188
	11	1	74	112	147	177	266	216
	12		79	118	154	185	209	225
	Daily Totals		488	764	1022	1262	1458	1544
May 21	6	6	5	7	9	10	11	11
	7	5	12	13	13	13	16	28
	8	4	15	15	25	45	63	80
	9	3	16	33	60	86	109	128
	10	2	22	56	89	120	146	167
	11	1	34	72	108	141	170	193
	12		37	77	114	149	178	201
	Daily Totals		246	469	724	982	1218	1436
June 21	6	6	7	9	10	12	12	13
	7	5	13	14	14	15	15	23
	8	4	16	16	16	35	55	73
	9	3	18	19	47	74	98	119
	10	2	18	41	74	105	133	157
	11	1	19	56	92	126	156	181
	12		22	60	98	133	168	189
	Daily Totals		204	370	610	874	1126	1356

Date	Solar Time		Solar Radiation, Btus per hour per square foot					
	AM	PM	Latitude (North)					
			24	32	40	48	56	64
July 21	6	6	6	8	9	11	12	12
	7	5	13	14	14	14	15	28
	8	4	16	16	24	43	61	77
	9	3	18	31	58	83	106	124
	10	2	21	54	86	116	142	162
	11	1	32	69	104	137	165	187
	12		36	74	111	144	173	195
	Daily Totals		246	458	702	956	1186	1400
Aug 21	6	6	2	4	5	6	7	7
	7	5	11	12	12	20	28	35
	8	4	16	33	50	65	78	87
	9	3	39	65	89	110	126	138
	10	2	58	91	120	146	166	179
	11	1	71	107	140	169	191	205
	12		75	113	147	177	200	215
	Daily Totals		470	736	978	1208	1392	1522
Sept 21	7	5	26	30	32	31	28	21
	8	4	62	75	84	90	89	81
	9	3	93	114	132	143	147	141
	10	2	116	145	168	185	193	189
	11	1	131	164	192	212	223	220
	12		136	171	200	221	233	230
	Daily Totals		992	1226	1416	1596	1594	1532

Date	Solar Time		Solar Radiation, Btus per hour per square foot					
	AM	PM	Latitude (North)					
			24	32	40	48	56	64
Oct 21	7	5	42	32	16	1	0	0
	8	4	99	104	100	87	57	10
	9	3	138	153	160	154	138	90
	10	2	165	188	203	207	195	155
	11	1	182	209	229	237	230	295
	12		188	217	238	247	241	208
	Daily Totals		1442	1588	1654	1626	1480	1106
Nov 21	7	5	29	1	0	0	0	0
	8	4	129	111	81	22	0	0
	9	3	172	176	167	135	58	0
	10	2	204	217	219	201	148	21
	11	1	222	241	248	238	196	79
	12		228	249	258	250	211	100
	Daily Totals		1730	1742	1686	1442	1016	300
Dec 21	7	5	14	0	0	0	0	0
	8	4	130	107	56	0	0	0
	9	3	184	183	163	109	4	0
	10	2	217	226	221	190	103	0
	11	1	236	251	252	231	164	4
	12		243	259	263	244	182	17
	Daily Totals		1808	1794	1646	1304	722	24

Appendix 3

Maps of the Average Percentage of the Time the Sun is Shining

JANUARY

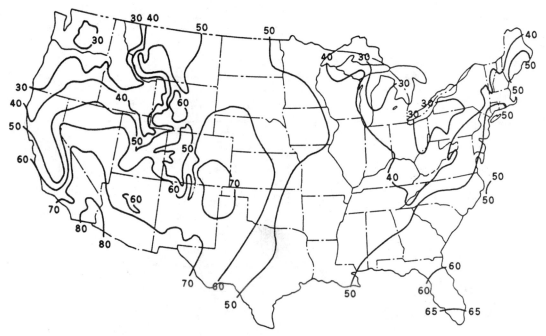

FEBRUARY

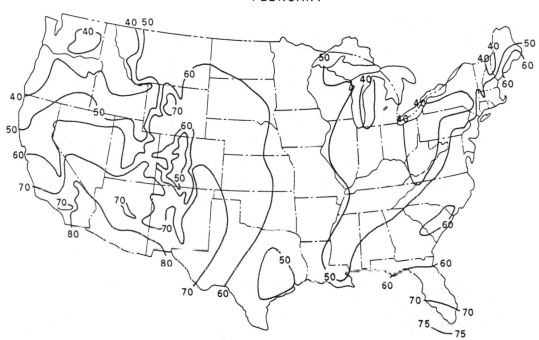

MARCH

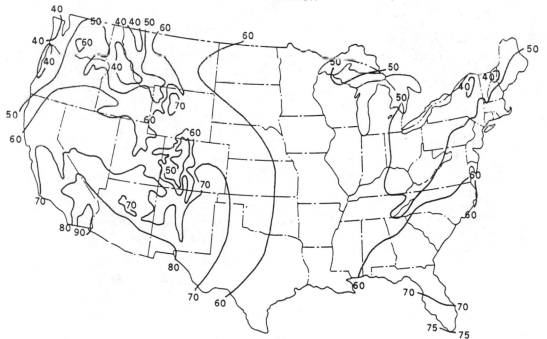

APRIL

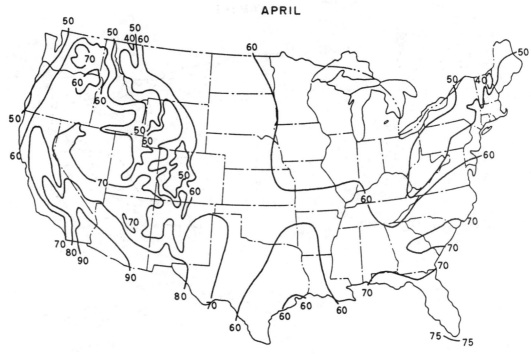

MAY

SEPTEMBER

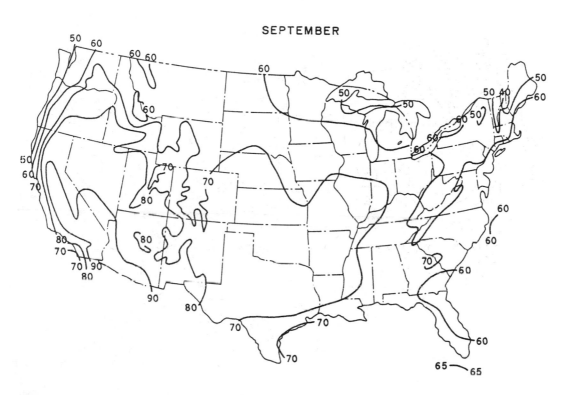

OCTOBER

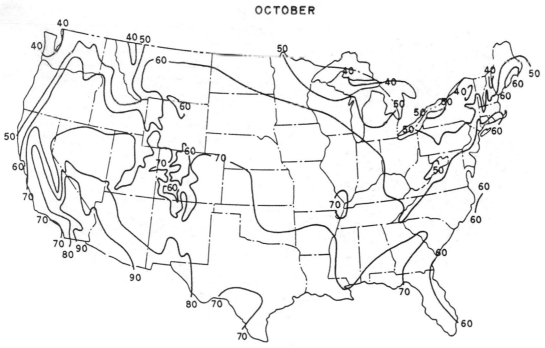

NOVEMBER

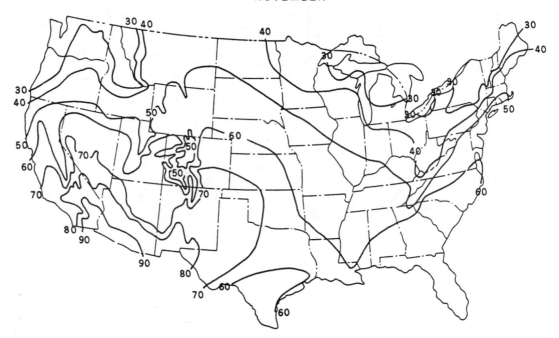

DECEMBER

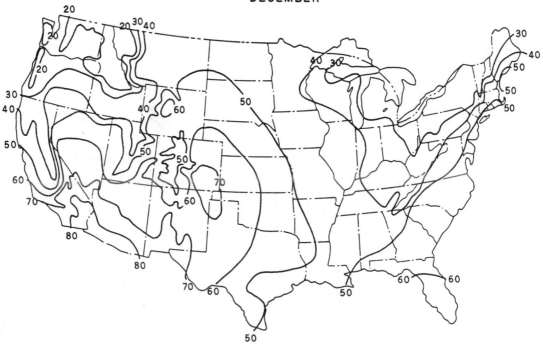

Appendix 4

Degree Days and Design Temperatures

State	City	Avg. Winter Temp	Design Temp	Sep	Oct	Nov	Dec	Jan	Feb	Mar	Apr	May	Yearly Total
Ala.	Birmingham	54.2	19	6	93	363	555	592	462	363	108	9	2551
	Huntsville	51.3	13	12	127	426	663	694	557	434	138	19	3070
	Mobile	59.9	26	0	22	213	357	415	300	211	42	0	1560
	Montgomery	55.4	22	0	68	330	527	543	417	316	90	0	2291
Alaska	Anchorage	23.0	−25	516	930	1284	1572	1631	1316	1293	879	592	10864
	Fairbanks	6.7	−53	642	1203	1833	2254	2359	1901	1739	1068	555	14279
	Juneau	32.1	− 7	483	725	921	1135	1237	1070	1073	810	601	9075
	Nome	13.1	−32	693	1094	1455	1820	1879	1666	1770	1314	930	14171
Ariz.	Flagstaff	35.6	0	201	558	867	1073	1169	991	911	651	437	7152
	Phoenix	58.5	31	0	22	234	415	474	328	217	75	0	1765
	Tucson	58.1	29	0	25	231	406	471	344	242	75	6	1800
	Winslow	43.0	9	6	245	711	1008	1054	770	601	291	96	4782
	Yuma	64.2	37	0	0	108	264	307	190	90	15	0	974
Ark.	Fort Smith	50.3	9	12	127	450	704	781	596	456	144	22	3292
	Little Rock	50.5	19	9	127	465	716	756	577	434	126	9	3219
	Texarkana	54.2	22	0	78	345	561	626	468	350	105	0	2533
Calif.	Bakersfield	55.4	31	0	37	282	502	546	364	267	105	19	2122
	Burbank	58.6	36	6	43	177	301	366	277	239	138	81	1646
	Eureka	49.9	32	258	329	414	499	546	470	505	438	372	4643
	Fresno	53.3	28	0	84	354	577	605	426	335	162	62	2611
	Long Beach	57.8	36	9	47	171	316	397	311	264	171	93	1803
	Los Angeles	57.4	41	42	78	180	291	372	302	288	219	158	2061
	Oakland	53.5	35	45	127	309	481	527	400	353	255	180	2870
	Sacramento	53.9	30	0	56	321	546	583	414	332	178	72	2502
	San Diego	59.5	42	21	43	135	236	298	235	214	135	90	1458
	San Francisco	55.1	42	102	118	231	388	443	336	319	279	239	3001
	Santa Maria	54.3	32	96	146	270	391	459	370	363	282	233	2967
Colo.	Alamosa	29.7	−17	279	639	1065	1420	1476	1162	1020	696	440	8529
	Colorado Springs	37.3	− 1	132	456	825	1032	1128	938	893	582	319	6423
	Denver	37.6	− 2	117	428	819	1035	1132	938	887	558	288	6283
	Grand Junction	39.3	8	30	313	786	1113	1209	907	729	387	146	5641
	Pueblo	40.4	− 5	54	326	750	986	1085	871	772	429	174	5462
Conn.	Bridgeport	39.9	4	66	307	615	986	1079	966	853	510	208	5617
	Hartford	37.3	1	117	394	714	1101	1190	1042	908	519	205	6235
	New Haven	39.0	5	87	347	648	1011	1097	991	871	543	245	5897
Del.	Wilmington	42.5	12	51	270	588	927	980	874	735	387	112	4930
D. C.	Washington	45.7	16	33	217	519	834	871	762	626	288	74	4224
Fla.	Daytona Beach	64.5	32	0	0	75	211	248	190	140	15	0	879
	Fort Myers	68.6	38	0	0	24	109	146	101	62	0	0	442
	Jacksonville	61.9	29	0	12	144	310	332	246	174	21	0	1239
	Key West	73.1	55	0	0	0	28	40	31	9	0	0	108
	Lakeland	66.7	35	0	0	57	164	195	146	99	0	0	661
	Miami	71.1	44	0	0	0	65	74	56	19	0	0	214
	Miami Beach	72.5	45	0	0	0	40	56	36	9	0	0	141
	Orlando	65.7	33	0	0	72	198	220	165	105	6	0	766
	Pensacola	60.4	29	0	19	195	353	400	277	183	36	0	1463
	Tallahassee	60.1	25	0	28	198	360	375	286	202	36	0	1485

State	City	Avg. Winter Temp	Design Temp	Sep	Oct	Nov	Dec	Jan	Feb	Mar	Apr	May	Yearly Total
	Tampa	66.4	36	0	0	60	171	202	148	102	0	0	683
	West Palm Beach	68.4	40	0	0	6	65	87	64	31	0	0	253
Ga.	Athens	51.8	17	12	115	405	632	642	529	431	141	22	2929
	Atlanta	51.7	18	18	124	417	648	636	518	428	147	25	2961
	Augusta	54.5	20	0	78	333	552	549	445	350	90	0	2397
	Columbus	54.8	23	0	87	333	543	552	434	338	96	0	2383
	Macon	56.2	23	0	71	297	502	505	403	295	63	0	2136
	Rome	49.9	16	24	161	474	701	710	577	468	177	34	3326
	Savannah	57.8	24	0	47	246	437	437	353	254	45	0	1819
Hawaii	Hilo	71.9	59	0	0	0	0	0	0	0	0	0	0
	Honolulu	74.2	60	0	0	0	0	0	0	0	0	0	0
Idaho	Boise	39.7	4	132	415	792	1017	1113	854	722	438	245	5809
	Lewiston	41.0	6	123	403	756	933	1063	815	694	426	239	5542
	Pocatello	34.8	− 8	172	493	900	1166	1324	1058	905	555	319	7033
Ill.	Chicago	37.5	− 4	81	326	753	1113	1209	1044	890	480	211	6155
	Moline	36.4	− 7	99	335	774	1181	1314	1100	918	450	189	6408
	Peoria	38.1	− 2	87	326	759	1113	1218	1025	849	426	183	6025
	Rockford	34.8	− 7	114	400	837	1221	1333	1137	961	516	236	6830
	Springfield	40.6	− 1	72	291	696	1023	1135	935	769	354	136	5429
Ind.	Evansville	45.0	6	66	220	606	896	955	767	620	237	68	4435
	Fort Wayne	37.3	0	105	378	783	1135	1178	1028	890	471	189	6205
	Indianapolis	39.6	0	90	316	723	1051	1113	949	809	432	177	5699
	South Bend	36.6	− 2	111	372	777	1125	1221	1070	933	525	239	6439
Iowa	Burlington	37.6	− 4	93	322	768	1135	1259	1042	859	426	177	6114
	Des Moines	35.5	− 7	96	363	828	1225	1370	1137	915	438	180	6568
	Dubuque	32.7	−11	156	450	906	1287	1420	1204	1026	546	260	7376
	Sioux City	34.0	−10	108	369	867	1240	1435	1198	989	483	214	6951
	Waterloo	32.6	−12	138	428	909	1296	1460	1221	1023	531	229	7320
Kans.	Dodge City	42.5	3	33	251	666	939	1051	840	719	354	124	4986
	Goodland	37.8	− 2	81	381	810	1073	1166	955	884	507	236	6141
	Topeka	41.7	3	57	270	672	980	1122	893	722	330	124	5182
	Wichita	44.2	5	33	229	618	905	1023	804	645	270	87	4620
Ky.	Covington	41.4	3	75	291	669	983	1035	893	756	390	149	5265
	Lexington	43.8	6	54	239	609	902	946	818	685	325	105	4683
	Louisville	44.0	8	54	248	609	890	930	818	682	315	105	4660
La.	Alexandria	57.5	25	0	56	273	431	471	361	260	69	0	1921
	Baton Rouge	59.8	25	0	31	216	369	409	294	208	33	0	1560
	Lake Charles	60.5	29	0	19	210	341	381	274	195	39	0	1459
	New Orleans	61.0	32	0	19	192	322	363	258	192	39	0	1385
	Shreveport	56.2	22	0	47	297	477	552	426	304	81	0	2184
Me.	Caribou	24.4	−18	336	682	1044	1535	1690	1470	1308	858	468	9767
	Portland	33.0	− 5	195	508	807	1215	1339	1182	1042	675	372	7511
Md.	Baltimore	43.7	12	48	264	585	905	936	820	679	327	90	4654
	Frederick	42.0	7	66	307	624	955	995	876	741	384	127	5087
Mass.	Boston	40.0	6	60	316	603	983	1088	972	846	513	208	5634
	Pittsfield	32.6	− 5	219	524	831	1231	1339	1196	1063	660	326	7578
	Worcester	34.7	− 3	147	450	774	1172	1271	1123	998	612	304	6969

State	City	Avg. Winter Temp	Design Temp	Sep	Oct	Nov	Dec	Jan	Feb	Mar	Apr	May	Yearly Total
Mich.	Alpena	29.7	− 5	273	580	912	1268	1404	1299	1218	777	446	8506
	Detroit	37.2	4	87	360	738	1088	1181	1058	936	522	220	6232
	Escanaba	29.6	− 7	243	539	924	1293	1445	1296	1203	777	456	8481
	Flint	33.1	− 1	159	465	843	1212	1330	1198	1066	639	319	7377
	Grand Rapids	34.9	2	135	434	804	1147	1259	1134	1011	579	279	6894
	Lansing	34.8	2	138	431	813	1163	1262	1142	1011	579	273	6909
	Marquette	30.2	− 8	240	527	936	1268	1411	1268	1187	771	468	8393
	Muskegon	36.0	4	120	400	762	1088	1209	1100	995	594	310	6696
	Sault Ste. Marie	27.7	−12	279	580	951	1367	1525	1380	1277	810	477	9048
Minn.	Duluth	23.4	−19	330	632	1131	1581	1745	1518	1355	840	490	10000
	Minneapolis	28.3	−14	189	505	1014	1454	1631	1380	1166	621	288	8382
	Rochester	28.8	−17	186	474	1005	1438	1593	1366	1150	630	301	8295
Miss.	Jackson	55.7	21	0	65	315	502	546	414	310	87	0	2239
	Meridian	55.4	20	0	81	339	518	543	417	310	81	0	2289
	Vicksburg	56.9	23	0	53	279	462	512	384	282	69	0	2041
Mo.	Columbia	42.3	2	54	251	651	967	1076	874	716	324	121	5046
	Kansas City	43.9	4	39	220	612	905	1032	818	682	294	109	4711
	St. Joseph	40.3	− 1	60	285	708	1039	1172	949	769	348	133	5484
	St. Louis	43.1	4	60	251	627	936	1026	848	704	312	121	4900
	Springfield	44.5	5	45	223	600	877	973	781	660	291	105	4900
Mont.	Billings	34.5	−10	186	487	897	1135	1296	1100	970	570	285	7049
	Glasgow	26.4	−25	270	608	1104	1466	1711	1439	1187	648	335	8996
	Great Falls	32.8	−20	258	543	921	1169	1349	1154	1063	642	384	7750
	Havre	28.1	−22	306	595	1065	1367	1584	1364	1181	657	338	8700
	Helena	31.1	−17	294	601	1002	1265	1438	1170	1042	651	381	8129
	Kalispell	31.4	− 7	321	654	1020	1240	1401	1134	1029	639	397	8191
	Miles City	31.2	−19	174	502	972	1296	1504	1252	1057	579	276	7723
	Missoula	31.5	− 7	303	651	1035	1287	1420	1120	970	621	391	8125
Neb.	Grand Island	36.0	− 6	108	381	834	1172	1314	1089	908	462	211	6530
	Lincoln	38.8	− 4	75	301	726	1066	1237	1016	834	402	171	5864
	Norfolk	34.0	−11	111	397	873	1234	1414	1179	983	498	233	6979
	North Platte	35.5	− 6	123	440	885	1166	1271	1039	930	519	248	6684
	Omaha	35.6	− 5	105	357	828	1175	1355	1126	939	465	208	6612
	Scottsbluff	35.9	− 8	138	459	876	1128	1231	1008	921	552	285	6673
Nev.	Elko	34.0	−13	225	561	924	1197	1314	1036	911	621	409	7433
	Ely	33.1	− 6	234	592	939	1184	1308	1075	977	672	456	7733
	Las Vegas	53.5	23	0	78	387	617	688	487	335	111	6	2709
	Reno	39.3	2	204	490	801	1026	1073	823	729	510	357	6332
	Winnemucca	36.7	1	210	536	876	1091	1172	916	837	573	363	6761
N. H.	Concord	33.0	−11	177	505	822	1240	1358	1184	1032	636	298	7383
N. J.	Atlantic City	43.2	14	39	251	549	880	936	848	741	420	133	4812
	Newark	42.8	11	30	248	573	921	983	876	729	381	118	4589
	Trenton	42.4	12	57	264	576	924	989	885	753	399	121	4980
N. M.	Albuquerque	45.0	14	12	229	642	868	930	703	595	288	81	4348
	Raton	38.1	− 2	126	431	825	1048	1116	904	834	543	301	6228
	Roswell	47.5	16	18	202	573	806	840	641	481	201	31	3793
	Silver City	48.0	14	6	183	525	729	791	605	518	261	87	3705

State	City	Avg. Winter Temp	Design Temp	Sep	Oct	Nov	Dec	Jan	Feb	Mar	Apr	May	Yearly Total
N. Y.	Albany	34.6	− 5	138	440	777	1194	1311	1156	992	564	239	6875
	Binghamton	36.6	− 2	141	406	732	1107	1190	1081	949	543	229	6451
	Buffalo	34.5	3	141	440	777	1156	1256	1145	1039	645	329	7062
	New York	42.8	11	30	233	540	902	986	885	760	408	118	4871
	Rochester	35.4	2	126	415	747	1125	1234	1123	1014	597	279	6748
	Schenectady	35.4	− 5	123	422	756	1159	1283	1131	970	543	211	6650
	Syracuse	35.2	− 2	132	415	744	1153	1271	1140	1004	570	248	6756
N. C.	Asheville	46.7	13	48	245	555	775	784	683	592	273	87	4042
	Charlotte	50.4	18	6	124	438	691	691	582	481	156	22	3191
	Greensboro	47.5	14	33	192	513	778	784	672	552	234	47	3805
	Raleigh	49.4	16	21	164	450	716	725	616	487	180	34	3393
	Wilmington	54.6	23	0	74	291	521	546	462	357	96	0	2347
	Winston-Salem	48.4	14	21	171	483	747	753	652	524	207	37	3595
N. D.	Bismarck	26.6	−24	222	577	1083	1463	1708	1442	1203	645	329	8851
	Devils Lake	22.4	−23	273	642	1191	1634	1872	1579	1345	753	381	9901
	Fargo	24.8	−22	219	574	1107	1569	1789	1520	1262	690	332	9226
	Williston	25.2	−21	261	601	1122	1513	1758	1473	1262	681	357	9243
Ohio	Akron-Canton	38.1	1	96	381	726	1070	1138	1016	871	489	202	6037
	Cincinnati	45.1	8	39	208	558	862	915	790	642	294	96	4410
	Cleveland	37.2	2	105	384	738	1088	1159	1047	918	552	260	6351
	Columbus	39.7	2	84	347	714	1039	1088	949	809	426	171	5660
	Dayton	39.8	0	78	310	696	1045	1097	955	809	429	167	5622
	Mansfield	36.9	1	114	397	768	1110	1169	1042	924	543	245	6403
	Toledo	36.4	1	117	406	792	1138	1200	1056	924	543	242	6494
	Youngstown	36.8	1	120	412	771	1104	1169	1047	921	540	248	6417
Okla.	Oklahoma City	48.3	11	15	164	498	766	868	664	527	189	34	3725
	Tulsa	47.7	12	18	158	522	787	893	683	539	213	47	3860
Ore.	Astoria	45.6	27	210	375	561	679	753	622	636	480	363	5186
	Eugene	45.6	22	129	366	585	719	803	627	589	426	279	4726
	Medford	43.2	21	78	372	678	871	918	697	642	432	242	5008
	Pendleton	42.6	3	111	350	711	884	1017	773	617	396	205	5127
	Portland	45.6	21	114	335	597	735	825	644	586	396	245	4635
	Roseburg	46.3	25	105	329	567	713	766	608	570	405	267	4491
	Salem	45.4	21	111	338	594	729	822	647	611	417	273	4754
Pa.	Allentown	38.9	3	90	353	693	1045	1116	1002	849	471	167	5810
	Erie	36.8	7	102	391	714	1063	1169	1081	973	585	288	6451
	Harrisburg	41.2	9	63	298	648	992	1045	907	766	396	124	5251
	Philadelphia	41.8	11	60	297	620	965	1016	889	747	392	118	5144
	Pittsburgh	38.4	5	105	375	726	1063	1119	1002	874	480	195	5987
	Reading	42.4	6	54	257	597	939	1001	885	735	372	105	4945
	Scranton	37.2	2	132	434	762	1104	1156	1028	893	498	195	6254
	Williamsport	38.5	1	111	375	717	1073	1122	1002	856	468	177	5934
R. I.	Providence	38.8	6	96	372	660	1023	1110	988	868	534	236	5954
S. C.	Charleston	57.9	26	0	34	210	425	443	367	273	42	0	1794
	Columbia	54.0	20	0	84	345	577	570	470	357	81	0	2484
	Florence	54.5	21	0	78	315	552	552	459	347	84	0	2387
	Greenville-Spartanburg	51.6	18	6	121	399	651	660	546	446	132	19	2980
S. D.	Huron	28.8	−16	165	508	1014	1432	1628	1355	1125	600	288	8223

State	City	Avg. Winter Temp	Design Temp	Sep	Oct	Nov	Dec	Jan	Feb	Mar	Apr	May	Yearly Total
	Rapid City	33.4	− 9	165	481	897	1172	1333	1145	1051	615	326	7345
	Sioux Falls	30.6	−14	168	462	972	1361	1544	1285	1082	573	270	7839
Tenn.	Bristol	46.2	11	51	236	573	828	828	700	598	261	68	4143
	Chattanooga	50.3	15	18	143	468	698	722	577	453	150	25	3254
	Knoxville	49.2	13	30	171	489	725	732	613	493	198	43	3494
	Memphis	50.5	17	18	130	447	698	729	585	456	147	22	3232
	Nashville	48.9	12	30	158	495	732	778	644	512	189	40	3578
Tex.	Abilene	53.9	17	0	99	366	586	642	470	347	114	0	2624
	Amarillo	47.0	8	18	205	570	797	877	664	546	252	56	3985
	Austin	59.1	25	0	31	225	388	468	325	223	51	0	1711
	Corpus Christi	64.6	32	0	0	120	220	291	174	109	0	0	914
	Dallas	55.3	19	0	62	321	524	601	440	319	90	6	2363
	El Paso	52.9	21	0	84	414	648	685	445	319	105	0	2700
	Galveston	62.2	32	0	6	147	276	360	263	189	33	0	1274
	Houston	61.0	28	0	6	183	307	384	288	192	36	0	1396
	Laredo	66.0	32	0	0	105	217	267	134	74	0	0	797
	Lubbock	48.8	11	18	174	513	744	800	613	484	201	31	3578
	Port Arthur	60.5	29	0	22	207	329	384	274	192	39	0	1447
	San Antonio	60.1	25	0	31	204	363	428	286	195	39	0	1546
	Waco	57.2	21	0	43	270	456	536	389	270	66	0	2030
	Wichita Falls	53.0	15	0	99	381	632	698	518	378	120	6	2832
Utah	Milford	36.5	− 1	99	443	867	1141	1252	988	822	519	279	6497
	Salt Lake City	38.4	5	81	419	849	1082	1172	910	763	459	233	6052
Vt.	Burlington	29.4	−12	207	539	891	1349	1513	1333	1187	714	353	8269
Va.	Lynchburg	46.0	15	51	223	540	822	849	731	605	267	78	4166
	Norfolk	49.2	20	0	136	408	698	738	655	533	216	37	3421
	Richmond	47.3	14	36	214	495	784	815	703	546	219	53	3865
	Roanoke	46.1	15	51	229	549	825	834	722	614	261	65	4150
Wash.	Olympia	44.2	21	198	422	636	753	834	675	645	450	307	5236
	Seattle	46.9	28	129	329	543	657	738	599	577	396	242	4424
	Spokane	36.5	− 2	168	493	879	1082	1231	980	834	531	288	6655
	Walla Walla	43.8	12	87	310	681	843	986	745	589	342	177	4805
	Yakima	39.1	6	144	450	928	1039	1163	868	713	435	220	5941
W. Va.	Charleston	44.8	9	63	254	591	865	880	770	648	300	96	4476
	Elkins	40.1	1	135	400	729	992	1008	896	791	444	198	5675
	Huntington	45.0	10	63	257	585	856	880	764	636	294	99	4446
	Parkersburg	43.5	8	60	264	606	905	942	826	691	339	115	4754
Wisc.	Green Bay	30.3	−12	174	484	924	1333	1494	1313	1141	654	335	8029
	La Crosse	31.5	−12	153	437	924	1339	1504	1277	1070	540	245	7589
	Madison	30.9	− 9	174	474	930	1330	1473	1274	1113	618	310	7863
	Milwaukee	32.6	− 6	174	471	876	1252	1376	1193	1054	642	372	7635
Wyo.	Casper	33.4	−11	192	524	942	1169	1290	1084	1020	657	381	7410
	Cheyenne	34.2	− 6	219	543	909	1085	1212	1042	1026	702	428	7381
	Lander	31.4	−16	204	555	1020	1299	1417	1145	1017	654	381	7870
	Sheridan	32.5	−12	219	539	948	1200	1355	1154	1051	642	366	7680

Directory of Resources

The purpose of this section is to help you obtain products and further information. Be informed before beginning any solar endeavor—from designing and building a house to organizaing an educational course or program. Rarely is it necessary to start from "scratch" in researching a solar project because a lot of free or inexpensive information already exists. We want to help you find what you need easily. Many direct sources are included, but to list all available solar sources would require many pages. Instead, we have provided major solar information centers and organizations. Our hope is that you will find this resource complete, yet fairly compact and easy to use.

State Energy Offices

Each state has an energy office that is, in many cases, one of the best sources for local information abou education materials, organizations, technical assistance, and reliable professional solar services. The following is a list of state energy offices.

Alabama Solar Energy Center
University of Alabama at Huntsville
Huntsville AL 35899
(205) 895-6361
(800)-228-5897 (in-state)

Alaska Housing Finance Corporation Energy
Resource & Information Center
520 E. 34th Avenue
Anchorage AK 99503-4199
(907) 564-9170
(800) 478-4638 (in-state, outside Anchorage)

Arizona Energy Office
Department of Commerce
3800 N. Central, 12th Floor
Phoenix AZ 85012
(602) 280-1402
(800) 392-8098 (in-state)

Arkansas Energy Office
One State Capital Mall
Little Rock AR 72201
(501) 682-7377

California
1516 Ninth Street, MS-25
Sacramento CA 95814-5512
(916)-654-4064
(800) 772-3300 (in-state)

Colorado Office of Energy Conservation
1675 Broadway, Suite1300
Denver CO 80203
(303) 620-4292
(800) 632-6662 (in-state)
303-620-4284 (hot line)

Connecticut Policy Development & Planning
Division
Energy Division
80 Washington Street
Hartford CT 06106-4459
(203) 566-2800

Delaware Division of Facilities
Management/Energy
P.O. Box 1401
Dover DE 19903
(302) 739-5644
(800) 282-8616 (in-state)

District of Columbia Energy Office
613 G Street NW, Suite 500
Washington DC 20001
(202) 727-1800

Florida Governor's Energy Office
2740 Centerview Drive
Tallahassee FL 32399-2100
(904) 488-6764

Georgia Office of Energy Resources
254 Washington SW, Suite 401
Atlanta GA 30334-8502
(404) 656-5176

Hawaii
Department of Business/Economic
Development
Energy Division
335 Merchant Street, Room 110
Honolulu HI 96813
(808) 587-3812

Idaho Department of Water Resources
Statehouse
1301 N. Orchard
Boise ID 83720
(208) 327-7959
(800) 334-7283 (in-state)

Illinois Department of Energy & Natural
Resources
325 W Adams, Room 300
Springfield IL 62704-1892
(217) 785-3969
(800) 252-8955 (in-state clearinghouse)

Indiana Office of Energy Policy
Indiana Department of Commerce
One North Capitol, Suite 700
Indianapolis IN 46204-2288
(317) 232-8940
(800) 382-4631 (in-state energy hot line)

Iowa Energy Bureau
Department of Natural Resources
Wallace Building
900 East Grand
Des Moines IA 50319-0034
(515) 281-7015

Kansas Department of Commerce & Housing
Division of Housing
Weatherization Program
700 SW Harrison St., Suite 1300
Topeka KS 66603-3712
(913) 296-2686

Kentucky Division of Energy
691 Teton Trail
Frankfurt KY 40601
(502) 564-7192

Louisiana Department of Natural Resources
Energy Division
P.O. Box 44156
Baton Rouge LA 70804-4156
(504) 342-1298

Maine Energy Conservation Division
Office of Community Development
219 Capitol Street, Station 130
Augusta ME 04333
(207) 289-6026

Maryland Energy Office
251 West Street
Annapolis MD 21401
(401) 626-1910

Massachusetts Department of Energy Resources
100 Cambridge St., 15th Floor
Boston MA 02202
(617) 727-4732

Michigan Public Service Commission
6545 Mercantile Way
P.O. Box 30221
Lansing MI 48909
(517) 334-6422

Minnesota Department of Public Service
Energy Division
150 E. Kellogg Boulevard
St. Paul MN 55101-1496
(800) 652-9747 (in-state)

Mississippi Energy Division
510 George Street
Jackson MS 39202
(610) 359-6600

Missouri Department of Natural Resources
Division of Energy
P.O. Box 176
Jefferson City MO 65102
(314) 751-4000

Montana Department of Natural Resources and
Conservation
1520 East Sixth Avenue
Helena MT 59620
(406) 444-6697

Nebraska State Energy Office
1200 N Street, Suite 10
P.O. Box 95085
Lincoln NE 68508
(402) 471-2867

Nevada Energy Office
Governor's Office of Community Services
400 W. King Street, Suite 400

Carson City NV 89710
(702) 687-4990

New Hampshire Office of Energy & Community
Services
57 Regional Drive
Concord NH 03301
(603) 271-2711
(800) 852-3466 (in-state)

New Jersey Board of Public Utilities
Office of the Secretary/Customer Assistance
Two Gateway Center
Newark NJ 07102
(201) 622-6103
(800) 492-4242 (in-state)

New Mexico Energy & Minerals Department
2040 S. Pacheco
Santa Fe NM 87505
(505) 827-5950

New York State Energy Office
Agency Building 2
Empire State Plaza
Albany NY 12223
(518) 473-0729
(800) 423-7283 (in-state)

North Carolina Solar Center
Box 7401
North Carolina State University
Raleigh NC 27695-7401
(919) 515-3799
(800)-662-7131 (in-state)

North Dakota Office of Intergovernmental
Assistance
State Capitol, 14th Floor
600 E. Boulevard Avenue
Bismarck ND 58505-0170
(701) 224-2094

Ohio Office of Energy Efficiency
Department of Development
77 S. High Street
Columbus OH 43266
(614) 466-6797

Oklahoma Department of Commerce
Division of Community Affairs and
Development
P.O. Box 26980
Oklahoma City OK 73126-0980
(405) 841-9321

Oregon Department of Energy
625 Marion Street, NE
Salem OR 97310
(503) 378-8446
(800) 221-8035 (in-state)

Pennsylvania Energy Office
116 Pine Street
Harrisburg PA 17101-1227
(717) 783-9981
(800) 682-7312 (in-state)

Rhode Island Office of Housing, Energy and
Intergovernmental Relations
275 Westminster Street, Suite 321
Providence RI 02903
(401) 277-3370

South Carolina Governor's Office of Energy
Program
1205 Pendleton Street, Suite 321
Columbia SC 29201
(803) 734-0325
(800) 851-8899 (in-state)

South Dakota Governor's Office of Energy
Policy
217 W. Missouri, Suite 200
Pierre SD 57501-4516
(605) 773-3603

Tennessee Department of Economic and
Community Development
Energy Division
320 Sixth Avenue North, 6th Floor
Nashville TN 37243-0405
(615) 741-6671
(800) 342-1340 (in-state)

Texas Governor's Energy Management Center
P.O. Box 12428
Austin TX 78711
(512) 463-1871

Utah Energy Office
3 Triad Center, Suite 450
Salt Lake City UT 84180-1204
(801) 538-5428
(800) 662-3633 (in-state)

Vermont Department of Public Service
Energy Efficiency Division
State Office Building
Montpelier VT 05620
(802) 828-2393
(800) 642-3281

Virginia Department of Mines, Minerals &
Energy
2201 W. Broad Street
Richmond VA 23220
(804) 367-6883

Washington State Energy Office
925 Plum Street
Town Square Building 4
Olympia WA 98504-3165
(206) 956-2000

West Virginia Division of Environmental
Protection
10 McJunkin Road
Nitro WV 25143-2506
(304) 759-0515
(800) 642-9012 (in-state)

Wisconsin Division of Energy &
Intergovernmental Relations
P.O. Box 7868
Madison WI 53707
(608) 266-1067

Wyoming Division of Economic and
Community Development
Energy Section
Derritt Building, 4th Floor N.
2301 Central
Cheyenne WY 82002
(307) 777-7284

National Organizations

Anyone seriously interested in solar energy will
stay abreast of the activities of these
organizations and obtain information about the
publications they offer.

Alliance to Save Energy
1725 K Street, NW, Suite 509
Washington DC 20006
(202) 857-0666

Alliance to Save Energy is a non-profit coalition
of business, government, environment and
consumer leaders. Their major goal is the
increased efficiency of energy use and so they
distribute low-cost or no-cost fact sheets and
pamphlets on financing energy efficiency at
home.

American Council for an Energy-Efficient
Economy

1001 Connecticut Avenue, NW, Suite 801
Washington DC 20036
(202) 429-8873

American Council for an Energy-Efficient
Economy (ACEEE) is a non-profit, non-
membership organization. They gather, evaluate
and disseminate information to stimulate greater
energy efficiency, specifically home appliances.
They conduct studies, publish books and
reports, including *Home Energy Magazine* and a
book, *Consumer Guide to Home Energy Savings.*
They provide expert testimony, and they
organize conferences.

American Solar Energy Society, Inc.
2400 Central Ave., Suite G-1
Boulder CO 80301
(303) 443-3130

American Solar Energy Society, Inc. (ASES),
founded in 1972, is a national organization with
state and regional chapters that is dedicated to
advancing the use of solar energy for the benefit
of U.S. citizens and the global environment.
ASES is affiliated with the International Solar
Energy Society and seeks to promote the
widespread near-term and long-term use of
solar energy. Providing a forum for the
exchange of information on solar energy
applications and research, ASES publishes *Solar
Today* six times per year. ASES chapters can be a
rich source of information and services. You
should get to know your local association and
the professional help and published materials
(including newsletters) and audiovisuals they
offer:

Alabama Solar Energy Association, c/o
UAH/ASEC, Solar Test Facility, Huntsville AL
35899. Contact: Robert E. Quick

Arizona Solar Energy Association, PO Box 1886, Chino Valley AZ 86323. (602) 636-2765. Contact: Mike Frerking

Florida Solar Energy Association, 9509 My Way Lane, NW, Ft. Myers FL 33919. (813) 489-2793

Illinois Solar Energy Association, 1361 Westchester, Glendale Heights IL 60139. (312) 690-3975. Contact: Richard Lewandowski

Kansas Solar Energy Society, 1602 S. McLean, Witchita KS 67217. Contact: Joseph T. Pajor

Minnesota Solar Energy Association, P.O. Box 65313, St. Paul MN 55165-0313. (612) 297-3067. Contact: Paul Helgeson

Mississippi Solar Energy Association, 225 W. Lumpkin Rd., Starkville MS 37959. (601) 323-7246. Contact: Dr. Pablo Okhuysen

Nebraska Solar Energy Association, University of Nebraska at Omaha, Engineering Building 110, 60th and Dodge, Omaha NE 68182. (402) 554-2669. Contact: John Thorp

New Mexico Solar Energy Association, P.O. Box 8507, Santa Fe NM 87504

North Carolina Solar Energy Association, 850 W. Morgan St., Raleigh NC 27603. (919) 832-7601. President: Ralph Cooke

Northeast Sustainable Energy Association, 23 Ames St., Greenfield MA 01301. (413) 774-6051. Contact: Nancy Hazard

Northern California Solar Energy Association, P.O. Box 3008, Berkeley CA 94703. Contact: Robin Mitchell

Ohio Solar Energy Association, Kent State University, School of Architecture, 304 Taylor Hall, Kent OH 44242. (216) 672-2869. Contact: Jack Kremers

Solar Energy Association of Oregon, 2637 SW Water Ave., Portland OR 97201. (503) 224-7867. Executive Director: Scott Lawrie

Southern California Energy Association, 339 20th St., Santa Monica CA 90402. Contact: Jack Cherne

Tennessee Solar Energy Association, Box 279, MTSU, Murfresboro TN 37132. (615) 898-2113. Contact: William F. Mathis

Texas Solar Energy Society, P.O. Box 14541, Austin TX 78761-4561. (512) 339-8562. Executive Director: Russel Smith

Center for Resourceful Building Technology
P.O. Box 3866
Missoula MT 59806
(406) 549-7678

Center for Resourceful Building Technology (CRBT) was established in 1990 to foster efficient energy and resource use within the building industry. CRBT conducts research, coordinates demonstration projects, and provides information on resource efficient technologies and materials to home builders, architects, and consumers through publications and lectures.

Conservation and Renewable Energy Inquiry and Referral Service (CAREIRS)
P.O. Box 8900
Silver Spring MD 20907
(800) 523-2929 (800-233-3071 in Alaska and Hawaii)

Conservation and Renewable Energy Inquiry and Referral Service (CAREIRS) provides the American public with residential and small business with information about energy conservation and renewable energy technologies, including—but not limited to—active and passive solar heating, photovoltaics, wind energy, biomass conversion, solar thermal electric, geothermal energy, small-scale hydroelectric, alcohol fuels, wood heating, ocean energy and all energy conservation technologies.

Energy Efficient Building Association (EEBA)
100 W. Campus Drive
Wausau WI 554-1
(715) 675-6331

Energy Efficient Building Association (EEBA) is a non-profit professional and educational association located at Northcentral Technical College in Wausau, Wisconsin. They provide information on the design, construction and operation of energy efficient buildings. EEBA sponsors conferences, workshops and expositions, and an annual Building Design Competition. They distribute publications on energy-efficient house design and construction.

Energy Rated Homes of America
100 Main Street, Suite 404
Little Rock AR 72201
(501) 374-7827

Energy Rated Homes of America is a national non-profit organization that rates the efficiency of homes.

Florida Solar Energy Center
300 State Road, Suite 401
Cape Canaveral FL 32920
(407) 783-0300

Florida Solar Energy Center (FSEC) conducts research on solar buildings, solar water heating systems, photovoltaics, and advanced technologies. Technical and consumer publications are available and workshops are offered.

Intelligent Buildings Institute
2101 L Street, Suite 300
Washington DC 20037

Intelligent Buildings Institute (IBI) is an international trade association established in 1986 to serve the needs of all sectors involved in advanced technologies used in commercial, institutional, and industrial buildings. IBI promotes the concept of intelligent buildings in the private marketplace, trade press, and with government agencies. IBI membership includes corporations, design professionals, and research institutes from 20 nations around the world and members are listed in a Directory of Products and Services.

National Association of Home Builders
1201 15th Street, NW
Washington DC 20005
(202) 822-0200

National Association of Home Builders (NAHB) is the trade association of the building industry. Members include small, moderate and large volume home builders, multi-family and commercial builders, and remodelers. Also subcontractors, retail and wholesale dealers, architects, engineers, lenders and realtors. State and local associations represent member interests on a local level.

National Appropriate Technology Assistance Service (NATAS)
P.O. Box 2525
Butte MT 59702
(800) 428-2525
(800) 428-1718 (in-state)

National Appropriate Technology Assistance
Service (NATAS) is a free service operated by
the National Center for Appropriate Technology
(NCAT), a non-profit corporation formed in
1976 to help people implement projects that use
renewable energy and energy-efficiency
technologies. Since 1984, NCAT has operated
NATAS under a contract with the U.S.
Department of Energy. NATAS provides
technical engineering assistance to help with
system design, component comparisons, system
problem solving, economic analysis, and
identifying local sources of assistance and
commercialization assistance with industry
overviews, microeconomic analyses, and
helping to identify and evaluate market trends
and potential funding sources.

National Renewable Energy Laboratory
1617 Cole Boulevard
Boulder CO 80401
(303) 231-7303

National Renewable Energy Laboratory (NREL),
formerly Solar Energy Research Institute (SERI),
is the nation's primary federal laboratory for
renewable energy research. It performs
functions assigned by DOE in research,
development, and testing. Their mission is to
facilitate the transfer of its renewable energy and
energy efficiency technologies to private
industry for commercialization.

Passive Solar Industries Council
1511 K Street NW, Suite 600
Washington DC 20005
(202) 628-7400

Passive Solar Industries Council (PSIC) is a
national association which provides information
on passive solar design and construction to the
U.S. building industry. Publications include

*Passive Solar Design Strategies: Guidelines for Home
Builders* and *Passive Solar Design Strategies:
Remodeling Guidelines for Conserving Energy at
Home.*

Renew America
1400 16th St., NW, Suite 710
Washington DC 20036
(202) 232-2252

Renew America, formerly called "Solar Lobby,"
is a non-profit organization specializing in
broad-based environmental program evaluation
and the promotion of positive initiatives for
change. By seeking out and promoting
successful programs, it offers positive,
constructive models to help communities meet
environmental challenges.

Rocky Mountain Institute
1739 Snowmass Creek Road
Snowmass CO 81654
(303) 927-3851

Rocky Mountain Institute (RMI) is a non-profit
research and educational foundation working
for the efficient and sustainable use of resources
as a path to global security. Founded in 1982 by
Amory and Hunter Lovins, RMI is best known
for its work on energy, but is also respected for
their important work with water, economic
renewal, agriculture, and security issues. In
1993, RMI was honored with a MacArthur
Foundation award.

Safe Energy Communication Council
1717 Massachusetts Avenue, NW, Suite LL215
Washington DC 20036
(202) 483-8491

SECC is a coalition of national environmental,
safe energy, and public interest media groups
working to increase public awareness of the

ability of energy efficiency and renewable energy sources to meet an increasing share of our nation's energy needs and the serious economic and environmental liabilities of nuclear power.

Small Homes Council-Building Research Council
University of Illinois
College of Fine and Applied Arts
One E. St. Mary's Rd
Champaign IL 61820
(217) 333-1801

Small Homes Council-Building Research Council was established by the University of Illinois as an agency for research publication, education, and public service in the area of housing and building. Efficient space utilization and energy conservation are key concerns. The Council does not draw or review house plans for individuals but they will try to answer questions relating to planning and construction not covered in their publications, preferably by mail.

Solar Energy Industries Association
777 No. Capital St., NE, Suite 805
Washington DC 20002-4226
(202) 408-0660

Solar Energy Industries Association (SEIA) is the national trade association for photovoltaic, passive, and solar thermal manufacturers, distributors, component suppliers, and contractors in the U.S. They publish the *Solar Industry Journal* quarterly.

Solar Rating & Certification Corporation
777 North Capitol Street, NE, Suite 805
Washington DC
(202) 408-0665

Solar Rating & Certification Corporation is a third party entity that rates equipment for solar hot water and solar pool heating systems.

Sustainable Technologies International
P.O. Box 1115
Carbonbdale CO 81623
(303) 963-0715

Sustainable Technologies International, formerly Solar Technology Institute, is a nonprofit corporation that provides education for renewable energy technologies and facilitates the use of these technologies in projects in less developed countries.

U.S. Department of Energy
1000 Independence Avenue, SW
Washington DC 20585
(202) 586-5000

U.S. Export Council for Renewable Energy
777 North Capitol Street, NW, Suite 805
Washington DC 20002
(202) 408-0665

U.S. Export Council for Renewable Energy is a consortium of seven U.S. trade associations that promotes the use of biomass, energy efficiency, geothermal, small-scale hydropower, photovoltaics, solar thermal, and wind energy technologies. Their primary objectives of market research, conferences, trade missions, and technical training encourage sustainable development in developing countries by facilitating the export of U.S. renewable energy technologies.

Passive Construction Plans

The following list of sources of energy efficient and solar home plans is from the National Appropriate Technology Assistance Service in Butte, Montana. NATAS is operated by the National Center for Appropriate Technology under contract to the U.S. Department of Energy.

Carolinas Concrete Masonry Association
1 Centerview Drive, Suite 112
Greensboro NC 27407
(919) 852-2074

Drawing-Room Graphic Services, Ltd.
Box 88627
North Vancouver BC Canada V7L 4L2
(604) 689-1841

Energetic Design, Inc.
P.O. Box 4446
Greensboro NC 27404
(919) 272-4660

Florida Solar Energy Center
300 State Road, #401
Cape Canaveral FL 32920
(407) 783-0300

Garlinhouse Company
P.O. Box 1717
Middletown CT 06457
(203) 632-0500 (in Connecticut)
(800) 235-5700 (outside Connecticut)

Home-Planners, Inc.
3275 W. Ina Road, Suite 110
Tucson AZ 85741
(800) 521-6797
(800) 322-6797
(602) 297-8200

HUD USER
P.O. BOX 6091
Rockville MD 20850
(800) 245-2691

Passive Solar Environments
821 W. Main Street
Kent OH 44240
(216) 673-7449

Small Homes Council
Building Research Council
University of Illinois-Urbana
No. 1 East St. Mary's Road
Champaign IL 61820
(217) 333-1801

Home Building Plan Services, Inc.
2235 NE Sandy Boulevard
Portland OR 97232
(503) 234-9337

Solar Home Plan Book
Northeast Utilities
Box 270
Hartford CT 06141
(203) 721-2715

Sunterra Homes, Inc.
132 NW Greenwood Avenue
Bend OR 97701
(503) 389-4733

Other house plan sources include:

American Ingenuity, Inc.
3500 Harlock Road
Melbourne FL 32935-7707
(407) 254-4220

Best Selling Home Plans and
Woman's Day Favorite Home Plans
Hachette Magazines
1633 Broadway
New York NY 10019
(800) 526-4667

Better Homes & Gardens Home Plan Ideas
Special Interest Publications
1716 Locust Street
Des Moines IA 50336

The Bloodgood Plan Service
3001 Grand Avenue
Des Moines IA 50312
(800) 752-6728

Home Building Plan Services, Inc.
2235 NE Sandy Boulevard
Portland OR 97232
(503) 234-9337

Passive Solar Environments
821 W. Main Street
Kent OH 44240
(216) 673-7449

The Garlinghouse Company, Inc.
34 Industrial Park Place
P.O. Box 1717
Middletown CT 06457
(800) 235-5700

Bibliography

Here are some of the best publications covering low-energy home design. The more you learn, the more satisfied you'll be when you use the sun. Some of the books and magazines are out of print, but the information they contain is invaluable. Check your local libraries and used bookstores.

Adams, Anthony. *Your Energy Efficient House: Building and Remodeling Ideas.* Charlotte, VT: Garden Way Publishing, 1975.

Adams, Jennifer A. *The Solar Church.* New York, NY: The Pilgrim Press, 1982.

AIA Research Corporation. "Passive Technology." *Research & Design Quarterly,* Vol 11, No. 3 (Fall 1979).

——————————. *A Survey of Passive Solar Buildings,* 1980, 1978; *Regional Guidelines for Building Passive Energy Conserving Homes,* 1978; *Solar Dwelling Design Concepts,* 1976. American Institute of Architects (Washington, D.C.)

——————————. *Solar Heating and Cooling Demonstration Program: A Descriptive Summary of: HUD Cycle 2 Solar Residential Projects,* 1976; *HUD Cycle 3 Solar Residential Projects,* 1977; *HUD Cycle 4 and 4a Solar Residential Projects,* 1977; *HUD Cycle 5 Solar Residential Projects,* 1980. U.S. Government Printing Office (Washington, D.C.)

All About Insulation. Lincoln, MA: Massachusetts Audubon Society.

Allen, Edward B. *How Buildings Work: The Natural Order of Architecture.* New York, NY: Oxford University Press, 1980.

Alward, Ron, and Andy Shapiro. *Low-Cost Passive Solar Greenhouses: A Design and Construction Guide.* New York, NY: Charles Scribners and Sons, 1980.

American Section of ISES. *Third National Passive Solar Conference.* Boulder, CO: American Section of International Solar Energy Society, 1979.

Ametek, Inc. *Solar Energy Handbook.* Radnor, PA: Chilton Book Co., 1983.

Anderson, Bruce. *Solar Energy: Fundamentals in Building Design.* New York, NY: McGraw-Hill, 1977.

——————————. *Solar Energy and Shelter Design.* Harrisville, NH: Total Environmental Action, 1973.

—————————— and J. Douglas Balcomb. *Passive Solar Design Handbook.* New York, NY: Van Nostrand Reinhold Co., 1984.

—————————— with Michael Riordan. *The New Solar Home Book.* Amherst, NH: Brick House Publishing Co., Inc., 1987.

ASES. *Proceedings of the National Passive Solar Conference: Annual Meetings.* Boulder, CO: American Solar Energy Society, Inc., 1977–present.

ASHRAE Handbook of Fundamentals. New York, NY: American Society of Heating, Refrigerating and Air Conditioning Engineers, 1967, 1972, 1977, 1982 and 1987.

Aulisi, Susan, and Doug McGilvray. *House Warming.* Edinburg, NY: Adirondack Alternate Energy, 1983.

Baer, Steve. *Sunspots*. Albuquerque, NM: Zomeworks Corp., 1975.

Bainbridge, David A. *The First Passive Solar Catalog*. Davis, CA: The Passive Solar Institute, 1978.

——————————. *The Integral Passive Solar Water Heater Book*. Davis, CA: The Passive Solar Institute, 1981.

——————————. *The Second Passive Solar Catalog*. Davis, CA: The Passive Solar Institute, 1980.

——————————, J. Corbett, and J. Hofacre. *Village Homes' Solar House Designs*. Emmaus, PA: Rodale Press, 1979.

Balcomb, J. Douglas. *Designing Passive Solar Buildings to Reduce Temperature Swings*. Los Alamos, NM: Los Alamos Scientific Laboratory, 1975.

——————————. *Energy Savings Obtainable Through Passive Solar Techniques*. Los Alamos, NM: Los Alamos Scientific Laboratory, 1978.

——————————, et al. *Passive Solar Heating Analysis: A Design Manual*. Atlanta, GA: ASHRAE, 1984.

——————————, and Robert W. Jones. *Workbook for Workshop on Advanced Passive Solar Design*. Santa Fe, NM: Balcolm Solar Associates, 1988.

Barnaby, C. S., P. Caesar, and B. Wilcox. *Solar for Your Present Home*. Sacramento, CA: California Energy Commission, 1977.

Bartholomew, Mel. *Square Foot Gardening*. Emmaus, PA: Rodale Press, 1981.

Beckett, H. E., and J. A. Godfrey. *Windows: Performance, Design, and Installation*. New York, NY: Van Nostrand Reinhold Co., 1974.

Beckman, W. A., S. A. Klein, and J. A. Duffie. *Solar Heating Design by the f-Chart Method*. New York, NY: John Wiley & Sons, 1977.

Benson, Tedd, with James Gruber. *Building the Timber Frame House*. Richmond, VT: Builders' Resource, 1990.

Berkeley Solar Group. *Passive Design Saves Energy and Money*. Sacramento, CA: Concrete Masonry Association of California and Nevada, 1981.

Berman, S. M., and S. D. Silverstein. *Energy Conservation and Window Systems*. Springfield, VA: National Technical Information Service, 1975.

Blandy, Thomas, and Denis Lamoureux. *All Through the House: A Guide to Home Weatherization*. New York, NY: McGraw-Hill Book Co., 1980.

Booth, Don, with Jonathan Booth and Peg Boyles. *Sun/Earth Buffering and Superinsulation*. Canterbury, NH: Community Builders, 1983.

Bowen, Arthur, et al. *Passive Cooling*. Boulder, CO: American Solar Energy Society, 1982.

Braden, Spruille III. *Graphic Standards of Solar Energy*. Boston, MA: CBI, 1977.

——————————— and Kathleen Steiner, with Alvin O'Konski. *Successful Solar Energy Solutions*. New York, NY: Van Nostrand Reinhold Co., 1980.

Buckley, Shawn. *Sun Up to Sun Down: Understanding Solar Energy*. New York, NY: McGraw-Hill Book Co., 1979.

Butti, Ken, and John Perlin. *A Golden Thread—500 Years of Solar Architecture.* Palo Alto, CA: Cheshire Books, 1980.

California Energy Commission. *Solar Gain: Passive Solar Design Competition.* Emmaus, PA: Rodale Press, 1980.

Campbell, Stu. *The Underground House Book.* Charlotte, VT: Garden Way Publishing, 1980.

Carter, C., and J. DeVilliers. *Principles of Passive Solar Building Design: With Microcomputer Applications.* Elmsford, NY: Pergamon Press, 1987.

Carter, Joe, ed. *Solarizing Your Present Home.* Emmaus, PA: Rodale Press, 1981.

Chandler, William, Marc Ledbetter, and Howard Geller. *Energy Efficiency: A New Agenda.* Washington, DC: American Council for an Energy-Efficient Economy, 1988.

Citizens' Advisory Committee on Environmental Quality. *Citizen Action Guide to Energy Conservation.* Washington, DC: Government Printing Office.

Claridge, David. "Window Management and Energy Savings." *Energy and Buildings,* 1, 19 (1977).

Clegg, P. and D. Watkins. *The Complete Greenhouse Book.* Charlotte, VT: Garden Way Publishing, 1978.

Cole, John N., and Charles Wing. *Breaking New Ground.* Boston, MA: Little, Brown and Co., 1986.

Cole, John N., and Charles Wing. *From the Ground Up: The Shelter Institute Guide to Building Your Own House.* Boston, MA: Atlantic Little, Brown & Co., 1976.

Colorado State University. *Solar Heating and Cooling of Residential Buildings: Sizing, Installation and Operation of Systems.* Washington, DC: U.S. Government Printing Office, 1980.

Conservation in Buildings: A Northwest Perspective. Butte, MT: NCAT, 1985.

Consumer Guide. *Energy Saver's Catalogue.* New York, NY: G.P Putnam's Sons, 1977.

Cook, Gary D. *A Guide to Residential Energy Efficiency in Florida.* Gainesville, FL: Cooperative Extension Service, University of Florida, 1990.

Cook, Jeffrey. *Award-Winning Passive Solar Designs: Professional Edition.* New York, NY: McGraw Hill, 1984.

——————. *Cool Houses for Desert Suburbs.* Phoenix, AZ: Arizona Solar Energy Commission, 1979.

——————. *Passive Cooling.* Cambridge, MA: The MIT Press, 1989.

—————— and Donald Prowler, eds. *Passive System Seventy-Eight.* Boulder, CO: American Solar Energy Society, 1978.

Crowley, Maureen, ed. *Energy: Sources of Print and Nonprint Materials.* New York, NY: Neal-Schuman Pubs., Inc., 1980.

Crowley, J. S., and L. Z. Zimmerman. *Practical Passive Solar Design: A Guide to Homebuilding & Land Development.* New York, NY: McGraw Hill, 1983.

Crowther, Richard L. *Affordable Passive Solar Homes: Low Cost Compact Designs.* Boulder, CO: American Solar Energy Society, 1983.

Crowther, Richard, and Solar Group/Architects. *Sun/Earth.* Denver, CO: A. D. Hirschfield Press, Inc., 1976.

Cutting Energy Costs. Washington, DC: U.S. Department of Agriculture, 1980.

Daniels, Farrington. *Direct Use of the Sun's Energy.* New Haven, CT: Yale University Press, 1983.

Daniels, G. *Solar Homes and Sun Heating.* New York, NY: Harper & Row, Inc., 1976.

DeKorne, J. and E. *The Survival Greenhouse—an Eco-System Approach to Home Food Production.* El Rito, NM: Walden Foundation, 1975.

Deryckx, Woody and Becky. *Two Solar Aquaculture-Greenhouse Systems for Western Washington: A Preliminary Report.* Seattle, WA: Tilth and Ecotope Group, Southfork Press, 1976.

Dubin, Fred, and G. Long. *Energy Conservation Standards.* New York, NY: McGraw-Hill Book Co., 1978.

duPont, Peter, and John Morrill. *Residential Indoor Air Quality and Energy Efficiency.* Washington, DC: American Council for an Energy-Efficient Economy, 1989.

Eccli, Eugene, ed. *Low-Cost, Energy Efficient Shelter for the Owner and Builder.* Emmaus, PA: Rodale Press, Inc., 1976.

Eklund, Ken, and David Baylon. *Design Tools for Energy Efficient Homes,* 3rd ed. Seattle, WA: Ecotope, 1984.

Energy Alternatives. Alexandria, VA: Time-Life Books, 1982.

Energy Efficient Building and Rebuilding: The Profit Opportunities. Boston, MA: Northeast Solar Energy Center, 1980.

Energy Conserving Features Inherent in Older Homes. Washington, DC: Department of Housing and Urban Development, 1982.

Energy-Efficient Construction Methods. Harrisville, NH: SolarVision Inc., 1982.

Energy Management Checklist for the Home. Washington, DC: U.S. Extension Service, 1983.

Energy-Wise Homebuyer: A Guide to Selecting an Energy-Efficient Home. Washington, DC: U.S. Department of Housing and Urban Development, 1979.

Financing Home Energy Improvements. Lincoln, MA: Massachusetts Audubon Society.

Find and Fix the Leaks: A Guide to Air Infiltration Reduction and Indoor Air Quality Control. Washington, DC: Department of Energy, 1981.

Fisher, Rick, and W. Yanda. *The Food and Heat Producing Solar Greenhouse: Design, Construction, Operation.* Santa Fe, NM: John Muir Publications, 1976.

Flavin, Christopher. *Energy and Architecture: The Solar & Conservation Potential.* Washington, DC: Worldwatch Institute, 1980.

Franta, Gregory, ed. *Solar Design Workbook.* Golden, CO: Solar Energy Research Institute, 1981.

Greenhouses for Living: The Complete Guide to Buying, Building, and Enjoying Residential

Sunspaces. New York, NY: Greenhouses for Living Information Center, 1989.

Greenhouse Gardening. Menlo Park, CA: Sunset Book, Lane Publishing, 1976.

Hayes, John, and Alex Wilson, eds. *Energy Conserving Solar Heated Greenhouses, Vol. III.* Boulder, CO: American Solar Energy Society, 1983.

Heat Saving Home Insulation. Harrisville, NH: SolarVision Inc., 1982.

Homeowner's Guide to Solar Heating. Menlo Park, CA: Sunset-Lane Publishing, 1979.

How to Weatherize Your Home or Apartment. Lincoln, MA: Massachusetts Audubon Society, 1986.

In the Bank—Or Up the Chimney? A Dollars and Cents Guide to Energy-Saving Home Improvements. Washington, DC: U.S. Dept. of Housing and Urban Development. 1977.

Jensen, Merle, ed. *Proceedings of the Solar Energy Food and Fuel Workshop* (April 1976). Tucson, AZ: University of Arizona, 1976.

Kadulski, Richard, and Terry Lyster. *Solplan 5: Energy Conserving Passive Solar Houses for Canada.* Vancouver, BC: The Drawing-Room Graphic Services Ltd., 1981.

Kern, Ken. *The Owner-Built Home.* Oakhurst, CA: Owner Builder Publications, Sierra Route, 1961.

Klein, Miriam, and Ron Alward. *Community Greenhouses Workbook: Where Do We Go From Here?* Boston, MA: Commonwealth of Massachusetts, Executive Office of Communities and Development.

Kreider, Jan F. . *The Solar Heating Design Process: Active and Passive Systems.* New York, NY: McGraw-Hill Book Co., 1982.

—————— and Frank Kreith. *Practical Design and Economics of Solar Heating and Cooling.* New York, NY: McGraw-Hill Book Co., 1975.

—————— and ——————. *Solar Energy Handbook.* New York, NY: McGraw-Hill Book Co., 1981.

—————— and ——————. *Solar Heating and Cooling: Active and Passive Design,* 2nd Edition. New York, NY: McGraw-Hill Book Co., 1975.

Langdon, Bill. *Movable Insulation: A Guide to Reducing Heating and Cooling Losses Through the Windows in Your Home.* Emmaus, PA: Rodale Press, Inc., 1983.

Leckie, Jim, et al. *Other Homes and Garbage.* San Francisco, CA: Sierra Club Books, 1975.

——————. *More Other Homes and Garbage.* San Francisco, CA: Sierra Club Books, 1981.

Lenchek, Thom, Chris Mattock, and John Raabe. *Superinsulated Design & Construction: A Guide to Building Energy-Efficient Homes.* Richmond, VT: Builders' Resource, 1991.

Levy, M. Emanuel, Deane Evans, and Cynthia Gardstein. *Passive Solar Construction Handbook.* Emmaus, Pa: Rodale Press, Inc., 1983.

Lodl, Kathleen A. *Solar Energy in Housing and Architecture: A Bibliography.* Monticello, IL: Van Bibliographies, 1987.

Lovins, Amory B. *Soft Energy Paths.* Cambridge, MA: Ballinger Publishing Co., 1977.

Lstiburek, Joseph. *Applied Building Science.* Downsview, Ontario: Building Engineering Corporation, 1987.

Magee, Tim, et al. *A Solar Greenhouse Guide for the Pacific Northwest,* 2nd ed. Seattle, WA: Ecotope, 1987.

Marbek Resource Consultants. *Air Sealing Homes for Energy Conservation.* Ottawa, Canada: Minister of Supply and Services Canada, 1984.

Marshall, Harold E., and Rosalie T. Ruegg. *Energy Conservation in Buildings: An Economics Guide for Investment Decisions.* Washington, DC: U.S. Government Printing Office, 1980.

Marshall, Brian and Robert Argue. *The Super-Insulated Retrofit Book: A Home Owner's Guide to Energy-Efficient Renovation.* Scarborough, Ontario: Firefly Books Ltd., 1981.

Mastalerz, John. *The Greenhouse Environment: The Effect of Environmental Factors on Flower Crops.* New York, NY: John Wiley and Sons Inc., 1977.

Mazria, Edward. *The Passive Solar Energy Book: Expanded Professional Edition.* Emmaus, PA: Rodale Press, Inc., 1979.

McCluney, W. R. *Window Treatment for Energy Conservation.* Cape Canaveral, FL: Florida Solar Energy Center, 1985.

McCullagh, James, ed. *The Solar Greenhouse Book.* Emmaus, PA: Rodale Press, Inc., 1978.

McPhillips, Martin., ed. *The Solar Age Resource Book.* New York, NY: Everest House, 1979.

Medinger, Larry. "Movable Insulation for Skylights." *Fine Homebuilding* (August/September 1985).

Meltzer, Michael. *Passive & Active Solar Heating Technology.* Englewood Cliffs, NJ: Prentice Hall, 1985.

Metz, Don. *Superhouse.* Charlotte, VT: Garden Way Publishing, 1981.

Mid-American Solar Energy Complex. *Solar 80 Home Designs.* Bloomington, MN: MASEC, 1980.

Moffit, Anne, and Marc Shiler. *Landscape Design That Saves Energy.* New York, NY: William Morrow & Co., Inc., 1981.

Moody, Robert J., John R. McBride, and MacDonald Homer. "Thermal Performance and Horticultural Production of an Attached Solar Greenhouse in a Severe Northern Climate." In *Proceedings of the Energy '82 Conference, Regina, Saskatchewan, Canada.* Butte, MT: NCAT, 1982.

Morris, Scott W. "Natural Convection Collectors." *Solar Age,* Vol. 3, No. 9 (September 1978)

Morrison, James W. *The Complete Energy-Saving Home Improvement Guide.* New York, NY: Arco Publishing Co., Inc. 1978.

National Association of Home Builders Research Foundation. *Insulation Manual.* Rockville, MD: National Association of Home Builders, 1979.

National Center for Appropriate Technology. *Energy Conservation and Renewable Energy Bibliography,* 1990; *Energy-Efficient Home Construction: Basic Superinsulation Techniques,* 1984; *Green-houses: Suggested Reading List,* 1990; *Low-Cost Passive Solar Greenhouses,* 1981; *Major*

Energy Conservation Retrofit, 1982; *Moisture and Home Energy Conservation: How to Detect, Solve and Avoid Related Problem*, 1983; *Solar Greenhouses and Sunspaces: Lessons Learned*, 1984; *Window Insulation: How to Sort Through the Options*, 1984. U.S. Government Printing Office (Washington, DC).

Nearing, Hell and Scott. *Building and Using our Sun-Heated Greenhouse*. Charlotte, VT: Garden Way Publishing, 1977.

Nelson, Gary. "Finding the Flaws in Superinsulation." *New England Builder* (August 1984).

New York State Energy Research and Development Authority. *1979 NYSERDA Passive Solar Design Awards*. Albany, NY, 1979.

————————————. *Making Your Own Solar Structures*. Albany, NY, 1982.

————————————. *Making Your Own Solar Wall Panel*. Albany, NY, 1982.

Nisson, J. D. Ned, and Gautam Dutt. *The Superinsulated Home Book*. New York, NY: John Wiley & Sons, 1985.

Northeast Solar Energy Center. *Proceedings of U.S. Department of Energy Passive and Hybrid Solar Energy Program Update Meeting*. Springfield, VA: National Technical Information Service, 1980.

Northeast Utilities. *Passive Solar Living*. Hartford, CT: Northeast Utilities, 1983.

————————————. *The Solar Home Planbook*, 2nd Edition. Hartford, CT: Northeast Utilities, 1985.

Olgyay, Aladar, and Victor Olgyay. *Solar Control and Shading Devices*. Princeton, NJ: Princeton University Press, 1967.

Passive Cooling Handbook. Berkeley, CA: Lawrence Berkeley Laboratory, 1980.

Passive Design Ideas for the Energy Conscious Builder. Philadelphia, PA: National Solar Heating and Cooling Information Center.

Passive Solar Heating Analysis: A Design Manual, 1984, and *Passive Solar Heating Analysis—Supplement One*, 1987, Atlanta, GA: American Heating Refrigerating and Air Conditioning Engineers.

Passive Solar Homes: A National Study. Washington, DC: U.S. Government Printing Office, 1986.

Passive Solar Performance and *Passive Solar Homes: 20 Case Studies*. Golden, CO: Solar Energy Research Institute, 1984.

Passive Solar Workshop Handbook. Santa Fe, NM: Passive Solar Associates, 1978.

Passive Solar Design: An Extensive Bibliography. Springfield, VA: National Technical Information Service, 1978.

Portland Cement Association. *Concrete Energy Conservation Guidelines*. Skokie, IL: Portland Cement Association, 1980.

Portola Institute. *Energy Primer*. Menlo Park, CA: Portola Institute, 1978.

Precast Concrete in Efficient Passive Solar Designs. Chicago, IL: Precast Prestressed Concrete Institute, 1979.

Price, Travis III. *Energy Conservation Guideline, Vol. I—New Construction, Energy Conservation Guideline, Vol. II—Rehabilitation* and *Energy Conservation Guideline, Vol. III—Effect of Occupant Behavior on Energy Use in an Inner City Neighborhood.* Pittsburg, PA: Carnegie-Mellon University, 1981.

Proceedings of the Solar Glazing 1979 Topical Conference. Philadelphia, PA: Mid-Atlantic Solar Energy Association, 1979.

Professional Builder. *Energy and the Builder.* Chicago, IL: Cahners Publishing Co., 1977.

Prowler, Douglas, ed. *Passive Solar State of the Art.* Boulder, CO: American Solar Energy Society, 1978.

Questions and Answers on Home Insulation. Washington, DC: Consumer Product Safety Commission, 1978.

Reif, Daniel K. *Solar Retrofit: Adding Solar to Your Home.* Amherst, NH: Brick House Publishing Co., 1981.

Reynolds, Smith, & Hills, Inc. *Life Cycle Costing Emphasizing Energy Conservation.* Springfield, VA. National Technical Information Service, 1976.

Robinette, Gary O., ed. *Landscape Planning for Energy Conservation.* Reston, VA: Environmental Design Press, 1977.

Rocky Mountain Institute. *Resource-Efficient Housing Guide.* Snowmass, CO: Rocky Mountain Institute, 1989.

Root, D., K. Sheinkopf, and C. Kettles. *Selling Solar Successfully.* Cape Canaveral, FL: Florida Solar Energy Center, 1985.

Rouse, Roland E. *Passive Solar Design for Multi-Family Buildings.* Danvers, MA: Bradford & Bigelow, 1982.

Rudoy, William, and Joseph F. Cuba, eds. *Cooling and Heating Load Calculation Manual.* Atlanta, GA: American Society of Heating, Refrigerating, and Air Conditioning Engineers, 1979.

Sabine, Hale J., and Myron B. Lacher. *Acoustical and Thermal Performance of Exterior Residential Walls.* Washington, DC: U.S. Government Printing Office, 1975.

Sandia Laboratories. *Passive Solar Buildings.* Springfield, VA: National Technical Information Service, 1979.

Save Money Save Energy. Washington, DC: Community Services Administration, 1977.

Schwolsky, Rick, and James I. Williams. *The Builder's Guide to Solar Construction.* New York, NY: McGraw-Hill, Inc., 1982.

Scully, Dan, D. Prowler, and Bruce Anderson. *The Fuel Savers.* Harrisville, NH: Total Environmental Action, Inc., 1975.

Selkowitz, Steven. *Windows for Energy Efficient Buildings,* Vol. 1, Nos. 1-2. Berkeley, CA: Lawrence Berkeley Laboratory, 1980.

Selling the Solar Home '80: Market Findings for the Housing Industry. Chicago, IL: Real Estate Research Corporation, 1980.

Shapiro, Andrew M. *The Homeowner's Complete Handbook for Add-On Solar Greenhouses & Sunspaces.* Emmaus, PA: Rodale Press, 1985.

Shurcliff, William A. *Solar Houses.* New York, NY: John Wiley and Sons, 1984.

——————————. *Thermal Shutters and Shades.* Amherst, NH: Brick House Publishing Co., 1980.

Skurka, Norma, and Jon Naar. *Design for a Limited Planet.* New York, NY: Ballantine Books, 1976.

Smith, Shane. *The Bountiful Solar Greenhouse.* Santa Fe, NM: John Muir Publications, 1982.

Sodha, M.S., et al. *Fundamentals of Solar Passive Building.* Elmsford, NY: Pergamon Press, 1986.

Solar Energy Research Institute. *Analysis Methods for Solar Heating and Cooling Applications.* Golden, CO: U.S. Department of Energy, 1980.

Solar Ideas for Your Home or Apartment. Lincoln, MA: Massachusetts Audubon Society, 1982.

Solar Systems Design. Harrisville, NH: SolarVision, Inc., 1983.

South Carolina Passive Solar Home Design: Low Cost Energy Efficient Opportunities. Columbia, SC: Governor's Office, Division of Energy Resources, 1980.

Southern Solar Energy Center. *Passive Retrofit Handbook.* Atlanta, GA: Southern Solar Energy Center, 1981.

Stein, Richard G. *Architecture and Energy.* New York, NY: Anchor Press/Doubleday, 1977.

Stephens, John Pierre, and Barbara Wezelman. *Movable Window Insulation.* Davis, CA: Sunwise Co-Op, 1982.

Sterling, Ray, et al. *Earth Sheltered Housing Design.* New York, NY: Van Nostrand Reinhold Co., 1979.

Steven Winter Associates. *Code Manual for Passive Solar Design—Single Family Residential Construction.* Atlanta, GA: Southern Solar Energy Center, 1979.

——————————. *Suntempering in the Northeast: A Selection of Builders' Designs.* Boston, MA: Northeast Solar Energy Center, 1980.

Stromberg, R. P., and S. O. Woodall. *Passive Solar Buildings: A Compilation of Data and Results.* Springfield, VA: National Technical Information Service, 1977.

Superinsulation: An Introduction to the Latest in Energy Efficient Construction. Lincoln, MA: Massachusetts Audubon Society, 1985.

Technical Aspects of Designing the Thermal Crafted™ Home. Toledo, OH: Owens-Corning Fiberglas Corporation, 1983.

Tennessee Valley Authority. *Solar Homes for the Valley: 1979 Design Portfolio.* Knoxville, TN: Tennessee Valley Authority, 1979.

——————————. *Solar Homes Design Portfolio.* Knoxville, TN: Tennessee Valley Authority, 1985.

Tips for Energy Savers. Washington, DC: U.S. Department of Energy, 1983.

TEA. *Solar Energy Home Design in Four Climates, 1975; The Thermal Mass Pattern Book, 1980;* Harrisville, NH: Total Environmental Action, Inc., 1975.

——. *Solar Home Heating in New Hampshire.* Concord, NH: N.H. Governor's Council on Energy, 1980.

Twitchell, Mary. *Solar Projects for Under $500.* Pownal, VT: Storey Communications, Inc., 1985.

U.S. Department of Housing and Urban Development Staff. *Passive Solar Homes.* Washington, DC: U.S. Government Printing Office, 1982.

Van Dresser, Peter. *Homegrown Sundwellings.* Sante Fe, NM: Lightning Tree, 1977.

Vance, Mary. *Solar Houses: A Bibliography.* Monticello, IL: Vance Bibliographies, 1989.

Vidich, Charles. *Passive Solar Subdivision Design: Tips for Developers, Builders and Site Planners.* Hartford, CT: Northeast Utilities, 1983.

——. *Protecting Solar Access: Tips for Builders and Homeowners.* Hartford, CT: Northeast Utilities, 1983.

——. *Regulating Passive Solar Subdivision Design: Tips for Planning and Zoning Commissions.* Hartford, CT: Northeast Utilities, 1983.

Viera, Robin K. and Kenneth G. Sheinkopf. *Energy-Efficient Florida Home Building.* Cape Canaveral, FL: Florida Solar Energy Center, 1988.

Vonier, Thomas. *Energy Conservation and Solar Energy for Historic Buildings: Guidelines for Appropriate Designs.* Washington, DC: National Center for Architecture and Urbanism, 1981.

Wade, Alex, and Neal Ewenstein. *30 Energy Efficient Houses You Can Build.* Emmaus, PA: Rodale Press, 1977.

Wade, Herbert, et al, eds. *Passive Solar: Subdivisions, Windows, Underground.* Boulder, CO: American Solar Energy Society, 1983.

Waldron, Wayne. "A Passive Solar Home Program for Florida." *Concrete Masonry Solar Architecture,* Vol 6, No. 3 (August 1986).

Watson, Donald, and Kenneth Labs. *Climate Design.* New York, NY: McGraw-Hill Book Co., 1983.

——. *Designing and Building a Solar House: Your Place in the Sun.* Charlotte, VT: Garden Way Publishing, 1977.

——, ed. *Energy Conservation through Building Design.* New York, NY: McGraw-Hill Book Co., 1979.

Wells, Malcolm. *Gentle Architecture.* Brewster, MA: Malcolm Wells, 1987.

Wilhelm, John L. "Solar Energy, the Ultimate Powerhouse." *National Geographic* (March 1976).

Wilson, Alex. *Thermal Storage Wall Design Manual.* Albuquerque, NM: Modern Press, 1979.

Wilson, Tom, ed. *Home Remedies: A Guidebook for Residential Retrofit.* Philadelphia, PA: Mid-Atlantic Solar Energy Association.

Windley, Leona. *Energy Efficient Interior Window Treatments.* Logan, UT: Utah State University, 1982.

Window Energy Systems. Harrisville, NH: SolarVision, Inc., 1983.

Wing, Charles. *From the Walls In.* Boston, MA: Little, Brown and Co., 1979.

——————, with John N. Cole. *From the Ground Up.* Boston, MA: Little, Brown and Co.

Wolf, Ray. *Insulating Window Shades.* Emmaus, PA: Rodale Press, 1980.

Wolfe, Delores E. *Growing Food in Solar Greenhouses.* New York, NY: Doubleday, 1981.

Wolpert, Cheri R. *Sun Rooms: Create a Beautiful Enclosed Glass Extension for Your Home.* Los Angeles, CA: Price Stern, 1989.

Wray, W. O., and J. D. Balcomb. *Sensitivity of Direct Gain Space Heating Performance to Fundamental Parameter Variations.* Los Alamos, NM: Los Alamos Scientific Laboratory, 1978.

Wright, David, and Dennis A. Andrejko. *Passive Solar Architecture: Logic & Beauty.* New York, NY: Van Nostrand Reinhold, 1982.

Yanda, William F., and Rick Fisher. *The Food and Heat Producing Solar Greenhouse.* Santa Fe, NM: John Muir Publications, Inc., 1976.

—————— and ——————. *Solar Heated Greenhouses.* Santa Fe, NM: John Muir Publications, 1980.

——————, and Susan Yanda. *An Attached Solar Greenhouse.* Santa Fe, NM: The Lightning Tree, 1976.

Yannas, S., ed. *Passive and Low Energy Architecture: Proceedings of the International Conference.* Elmsford, NY: Pergamon Press, 1983.